STUDENT SOLUTIONS M**A**

Neil Wigley
University of Windsor

to accompany

CALCULUS
Multivariable

Eighth Edition

Howard Anton
Drexel University

Irl C. Bivens
Davidson College

Stephen L. Davis
Davidson College

WILEY

JOHN WILEY & SONS, INC.

Cover Photo: ©Arthur Tilley/Taxi/Getty Images

To order books or for customer service, please call 1-800-CALL-WILEY (225-5945).

ISBN 0-471-67212-2

Printed in the United States of America

10 9 8 7 6 5 4 3 2

Printed and bound by Malloy Lithographing, Inc.

CONTENTS

CHAPTER 12
Three-Dimensional Space; Vectors

EXERCISE SET 12.1

1. **(a)** $(0,0,0),(3,0,0),(3,5,0),(0,5,0),(0,0,4),(3,0,4),(3,5,4),(0,5,4)$

(b) $(0,1,0),(4,1,0),(4,6,0),(0,6,0),(0,1,-2),(4,1,-2),(4,6,-2),(0,6,-2)$

3. corners: $(4,2,-2),\ (4,2,1),\ (4,1,1),\ (4,1,-2),$
$(-6,1,1),\ (-6,2,1),\ (-6,2,-2),\ (-6,1,-2)$

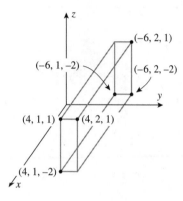

5. **(a)** a single point on that line   **(b)** a line in that plane   **(c)** a plane in 3−space

7. The diameter is $d = \sqrt{(1-3)^2 + (-2-4)^2 + (4+12)^2} = \sqrt{296}$, so the radius is $\sqrt{296}/2 = \sqrt{74}$. The midpoint $(2,1,-4)$ of the endpoints of the diameter is the center of the sphere.

9. **(a)** The sides have lengths 7, 14, and $7\sqrt{5}$; it is a right triangle because the sides satisfy the Pythagorean theorem, $(7\sqrt{5})^2 = 7^2 + 14^2$.

(b) $(2,1,6)$ is the vertex of the 90° angle because it is opposite the longest side (the hypotenuse).

(c) area $= (1/2)(\text{altitude})(\text{base}) = (1/2)(7)(14) = 49$

11. **(a)** $(x-1)^2 + y^2 + (z+1)^2 = 16$

(b) $r = \sqrt{(-1-0)^2 + (3-0)^2 + (2-0)^2} = \sqrt{14},\ (x+1)^2 + (y-3)^2 + (z-2)^2 = 14$

(c) $r = \dfrac{1}{2}\sqrt{(-1-0)^2 + (2-2)^2 + (1-3)^2} = \dfrac{1}{2}\sqrt{5}$, center $(-1/2, 2, 2)$,
$(x+1/2)^2 + (y-2)^2 + (z-2)^2 = 5/4$

13. $(x-2)^2 + (y+1)^2 + (z+3)^2 = r^2$,

(a) $r^2 = 3^2 = 9$ **(b)** $r^2 = 1^2 = 1$ **(c)** $r^2 = 2^2 = 4$

15. Let the center of the sphere be (a,b,c). The height of the center over the x-y plane is measured along the radius that is perpendicular to the plane. But this is the radius itself, so height = radius, i.e. $c = r$. Similarly $a = r$ and $b = r$.

17. $(x+5)^2 + (y+2)^2 + (z+1)^2 = 49$; sphere, $C(-5,-2,-1)$, $r = 7$

19. $(x-1/2)^2 + (y-3/4)^2 + (z+5/4)^2 = 54/16$; sphere, $C(1/2, 3/4, -5/4)$, $r = 3\sqrt{6}/4$

21. $(x - 3/2)^2 + (y + 2)^2 + (z - 4)^2 = -11/4$; no graph

23. **(a)** 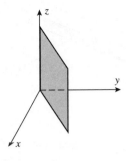   **(b)** 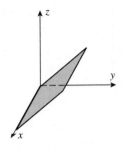   **(c)**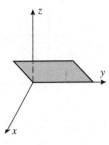

25. **(a)** 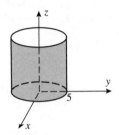   **(b)** 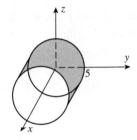   **(c)**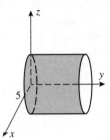

27. **(a)** $-2y + z = 0$ **(b)** $-2x + z = 0$
(c) $(x - 1)^2 + (y - 1)^2 = 1$ **(d)** $(x - 1)^2 + (z - 1)^2 = 1$

29. 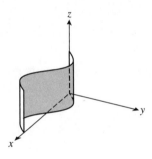   **31.**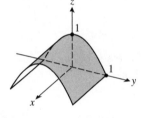

33. 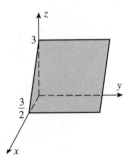   **35.**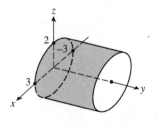

37.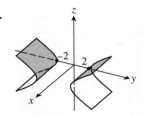

39. (a)

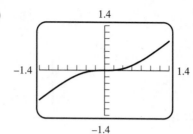

(b)

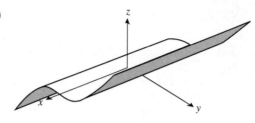

41. Complete the square to get $(x + 1)^2 + (y - 1)^2 + (z - 2)^2 = 9$; center $(-1, 1, 2)$, radius 3. The distance between the origin and the center is $\sqrt{6} < 3$ so the origin is inside the sphere. The largest distance is $3 + \sqrt{6}$, the smallest is $3 - \sqrt{6}$.

43. $(y + 3)^2 + (z - 2)^2 > 16$; all points outside the circular cylinder $(y + 3)^2 + (z - 2)^2 = 16$.

45. Let r be the radius of a styrofoam sphere. The distance from the origin to the center of the bowling ball is equal to the sum of the distance from the origin to the center of the styrofoam sphere nearest the origin and the distance between the center of this sphere and the center of the bowling ball so

$$\sqrt{3}R = \sqrt{3}r + r + R, \ (\sqrt{3} + 1)r = (\sqrt{3} - 1)R, \ r = \frac{\sqrt{3} - 1}{\sqrt{3} + 1}R = (2 - \sqrt{3})R.$$

47. (a) At $x = c$ the trace of the surface is the circle $y^2 + z^2 = [f(c)]^2$, so the surface is given by $y^2 + z^2 = [f(x)]^2$

(b) $y^2 + z^2 = e^{2x}$ **(c)** $y^2 + z^2 = 4 - \frac{3}{4}x^2$, so let $f(x) = \sqrt{4 - \frac{3}{4}x^2}$

49. $(a \sin \phi \cos \theta)^2 + (a \sin \phi \sin \theta)^2 + (a \cos \phi)^2 = a^2 \sin^2 \phi \cos^2 \theta + a^2 \sin^2 \phi \sin^2 \theta + a^2 \cos^2 \phi$

$$= a^2 \sin^2 \phi(\cos^2 \theta + \sin^2 \theta) + a^2 \cos^2 \phi$$

$$= a^2 \sin^2 \phi + a^2 \cos^2 \phi = a^2(\sin^2 \phi + \cos^2 \phi) = a^2$$

EXERCISE SET 12.2

1. (a–c)

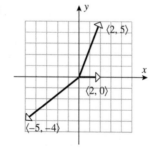

(d–f)

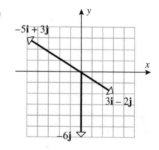

3. **(a–b)** 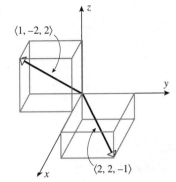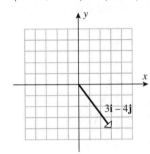   **(c–d)**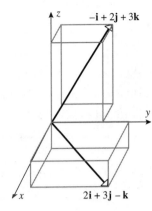

5. **(a)** $\langle 4 - 1, 1 - 5 \rangle = \langle 3, -4 \rangle$ **(b)** $\langle 0 - 2, 0 - 3, 4 - 0 \rangle = \langle -2, -3, 4 \rangle$

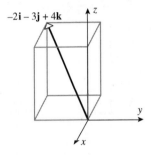

7. **(a)** $\langle 2 - 3, 8 - 5 \rangle = \langle -1, 3 \rangle$

   (b) $\langle 0 - 7, 0 - (-2) \rangle = \langle -7, 2 \rangle$

   (c) $\langle -3, 6, 1 \rangle$

9. **(a)** Let (x, y) be the terminal point, then $x - 1 = 3$, $x = 4$ and $y - (-2) = -2$, $y = -4$.
   The terminal point is $(4, -4)$.

   (b) Let (x, y, z) be the initial point, then $5 - x = -3$, $-y = 1$, and $-1 - z = 2$ so $x = 8$,
   $y = -1$, and $z = -3$. The initial point is $(8, -1, -3)$.

11. **(a)** $-\mathbf{i} + 4\mathbf{j} - 2\mathbf{k}$ **(b)** $18\mathbf{i} + 12\mathbf{j} - 6\mathbf{k}$ **(c)** $-\mathbf{i} - 5\mathbf{j} - 2\mathbf{k}$

   (d) $40\mathbf{i} - 4\mathbf{j} - 4\mathbf{k}$ **(e)** $-2\mathbf{i} - 16\mathbf{j} - 18\mathbf{k}$ **(f)** $-\mathbf{i} + 13\mathbf{j} - 2\mathbf{k}$

13. **(a)** $\|\mathbf{v}\| = \sqrt{1 + 1} = \sqrt{2}$ **(b)** $\|\mathbf{v}\| = \sqrt{1 + 49} = 5\sqrt{2}$

   (c) $\|\mathbf{v}\| = \sqrt{21}$ **(d)** $\|\mathbf{v}\| = \sqrt{14}$

15. **(a)** $\|\mathbf{u} + \mathbf{v}\| = \|2\mathbf{i} - 2\mathbf{j} + 2\mathbf{k}\| = 2\sqrt{3}$ **(b)** $\|\mathbf{u}\| + \|\mathbf{v}\| = \sqrt{14} + \sqrt{2}$

   (c) $\| -2\mathbf{u}\| + 2\|\mathbf{v}\| = 2\sqrt{14} + 2\sqrt{2}$ **(d)** $\|3\mathbf{u} - 5\mathbf{v} + \mathbf{w}\| = \| -12\mathbf{j} + 2\mathbf{k}\| = 2\sqrt{37}$

   (e) $(1/\sqrt{6})\mathbf{i} + (1/\sqrt{6})\mathbf{j} - (2/\sqrt{6})\mathbf{k}$ **(f)** 1

17. **(a)** $\| -\mathbf{i} + 4\mathbf{j}\| = \sqrt{17}$ so the required vector is $\left(-1/\sqrt{17}\right)\mathbf{i} + \left(4/\sqrt{17}\right)\mathbf{j}$

   (b) $\|6\mathbf{i} - 4\mathbf{j} + 2\mathbf{k}\| = 2\sqrt{14}$ so the required vector is $(-3\mathbf{i} + 2\mathbf{j} - \mathbf{k})/\sqrt{14}$

   (c) $\overrightarrow{AB} = 4\mathbf{i} + \mathbf{j} - \mathbf{k}$, $\|\overrightarrow{AB}\| = 3\sqrt{2}$ so the required vector is $(4\mathbf{i} + \mathbf{j} - \mathbf{k})/\left(3\sqrt{2}\right)$

19. (a) $-\dfrac{1}{2}\mathbf{v} = \langle -3/2, 2\rangle$ **(b)** $\|\mathbf{v}\| = \sqrt{85}$, so $\dfrac{\sqrt{17}}{\sqrt{85}}\mathbf{v} = \dfrac{1}{\sqrt{5}}\langle 7, 0, -6\rangle$ has length $\sqrt{17}$

21. (a) $\mathbf{v} = \|\mathbf{v}\|\langle\cos(\pi/4), \sin(\pi/4)\rangle = \langle 3\sqrt{2}/2, 3\sqrt{2}/2\rangle$

    (b) $\mathbf{v} = \|\mathbf{v}\|\langle\cos 90°, \sin 90°\rangle = \langle 0, 2\rangle$

    (c) $\mathbf{v} = \|\mathbf{v}\|\langle\cos 120°, \sin 120°\rangle = \langle -5/2, 5\sqrt{3}/2\rangle$

    (d) $\mathbf{v} = \|\mathbf{v}\|\langle\cos \pi, \sin \pi\rangle = \langle -1, 0\rangle$

23. From (12), $\mathbf{v} = \langle\cos 30°, \sin 30°\rangle = \langle\sqrt{3}/2, 1/2\rangle$ and $\mathbf{w} = \langle\cos 135°, \sin 135°\rangle = \langle -\sqrt{2}/2, \sqrt{2}/2\rangle$, so $\mathbf{v} + \mathbf{w} = ((\sqrt{3} - \sqrt{2})/2, (1 + \sqrt{2})/2)$

25. (a) The initial point of $\mathbf{u} + \mathbf{v} + \mathbf{w}$ is the origin and the endpoint is $(-2, 5)$, so $\mathbf{u} + \mathbf{v} + \mathbf{w} = \langle -2, 5\rangle$.

    (b) The initial point of $\mathbf{u} + \mathbf{v} + \mathbf{w}$ is $(-5, 4)$ and the endpoint is $(-2, -4)$, so $\mathbf{u} + \mathbf{v} + \mathbf{w} = \langle 3, -8\rangle$.

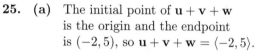
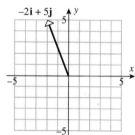
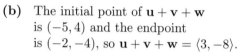
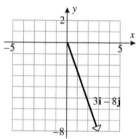

27. $6\mathbf{x} = 2\mathbf{u} - \mathbf{v} - \mathbf{w} = \langle -4, 6\rangle, \mathbf{x} = \langle -2/3, 1\rangle$

29. $\mathbf{u} = \dfrac{5}{7}\mathbf{i} + \dfrac{2}{7}\mathbf{j} + \dfrac{1}{7}\mathbf{k}, \mathbf{v} = \dfrac{8}{7}\mathbf{i} - \dfrac{1}{7}\mathbf{j} - \dfrac{4}{7}\mathbf{k}$

31. $\|(\mathbf{i} + \mathbf{j}) + (\mathbf{i} - 2\mathbf{j})\| = \|2\mathbf{i} - \mathbf{j}\| = \sqrt{5}, \|(\mathbf{i} + \mathbf{j}) - (\mathbf{i} - 2\mathbf{j})\| = \|3\mathbf{j}\| = 3$

33. (a) $5 = \|k\mathbf{v}\| = |k|\|\mathbf{v}\| = \pm 3k$, so $k = \pm 5/3$

    (b) $6 = \|k\mathbf{v}\| = |k|\|\mathbf{v}\| = 2\|\mathbf{v}\|$, so $\|\mathbf{v}\| = 3$

35. (a) Choose two points on the line, for example $P_1(0, 2)$ and $P_2(1, 5)$; then $\overrightarrow{P_1 P_2} = \langle 1, 3\rangle$ is parallel to the line, $\|\langle 1, 3\rangle\| = \sqrt{10}$, so $\langle 1/\sqrt{10}, 3/\sqrt{10}\rangle$ and $\langle -1/\sqrt{10}, -3/\sqrt{10}\rangle$ are unit vectors parallel to the line.

    (b) Choose two points on the line, for example $P_1(0, 4)$ and $P_2(1, 3)$; then $\overrightarrow{P_1 P_2} = \langle 1, -1\rangle$ is parallel to the line, $\|\langle 1, -1\rangle\| = \sqrt{2}$ so $\langle 1/\sqrt{2}, -1/\sqrt{2}\rangle$ and $\langle -1/\sqrt{2}, 1/\sqrt{2}\rangle$ are unit vectors parallel to the line.

    (c) Pick any line that is perpendicular to the line $y = -5x + 1$, for example $y = x/5$; then $P_1(0, 0)$ and $P_2(5, 1)$ are on the line, so $\overrightarrow{P_1 P_2} = \langle 5, 1\rangle$ is perpendicular to the line, so $\pm\dfrac{1}{\sqrt{26}}\langle 5, 1\rangle$ are unit vectors perpendicular to the line.

37. (a) the circle of radius 1 about the origin

    (b) the closed disk of radius 1 about the origin

    (c) all points outside the closed disk of radius 1 about the origin

39. (a) the (hollow) sphere of radius 1 about the origin

    (b) the closed ball of radius 1 about the origin

    (c) all points outside the closed ball of radius 1 about the origin

41. Since $\phi = \pi/2$, from (14) we get $\|\mathbf{F}_1 + \mathbf{F}_2\|^2 = \|\mathbf{F}_1\|^2 + \|\mathbf{F}_2\|^2 = 3600 + 900$,

so $\|\mathbf{F}_1 + \mathbf{F}_2\| = 30\sqrt{5}$ lb, and $\sin\alpha = \dfrac{\|\mathbf{F}_2\|}{\|\mathbf{F}_1 + \mathbf{F}_2\|}\sin\phi = \dfrac{30}{30\sqrt{5}}, \alpha \approx 26.57°, \theta = \alpha \approx 26.57°.$

43. $\|\mathbf{F}_1 + \mathbf{F}_2\|^2 = \|\mathbf{F}_1\|^2 + \|\mathbf{F}_2\|^2 + 2\|\mathbf{F}_1\|\|\mathbf{F}_2\|\cos\phi = 160{,}000 + 160{,}000 - 2(400)(400)\dfrac{\sqrt{3}}{2}$,

so $\|\mathbf{F}_1 + \mathbf{F}_2\| \approx 207.06$ N, and $\sin\alpha = \dfrac{\|\mathbf{F}_2\|}{\|\mathbf{F}_1 + \mathbf{F}_2\|}\sin\phi \approx \dfrac{400}{207.06}\left(\dfrac{1}{2}\right), \alpha = 75.00°,$

$\theta = \alpha - 30° = 45.00°.$

45. Let $\mathbf{F}_1, \mathbf{F}_2, \mathbf{F}_3$ be the forces in the diagram with magnitudes $40, 50, 75$ respectively. Then $\mathbf{F}_1 + \mathbf{F}_2 + \mathbf{F}_3 = (\mathbf{F}_1 + \mathbf{F}_2) + \mathbf{F}_3$. Following the examples, $\mathbf{F}_1 + \mathbf{F}_2$ has magnitude 45.83 N and makes an angle 79.11° with the positive x-axis. Then $\|(\mathbf{F}_1 + \mathbf{F}_2) + \mathbf{F}_3\|^2 \approx 45.83^2 + 75^2 + 2(45.83)(75)\cos 79.11°$, so $\mathbf{F}_1 + \mathbf{F}_2 + \mathbf{F}_3$ has magnitude ≈ 94.995 N and makes an angle $\theta = \alpha \approx 28.28°$ with the positive x-axis.

47. Let $\mathbf{F}_1, \mathbf{F}_2$ be the forces in the diagram with magnitudes $8, 10$ respectively. Then $\|\mathbf{F}_1 + \mathbf{F}_2\|$ has magnitude $\sqrt{8^2 + 10^2 + 2 \cdot 8 \cdot 10 \cos 120°} = 2\sqrt{21} \approx 9.165$ lb, and makes an angle $60° + \sin^{-1}\dfrac{\|\mathbf{F}_1\|}{\|\mathbf{F}_1 + \mathbf{F}_2\|}\sin 120 \approx 109.11°$ with the positive x-axis, so $\mathbf{F}$ has magnitude 9.165 lb and makes an angle $-70.89°$ with the positive x-axis.

49. $\mathbf{F}_1 + \mathbf{F}_2 + \mathbf{F} = \mathbf{0}$, where $\mathbf{F}$ has magnitude 250 and makes an angle $-90°$ with the positive x-axis. Thus $\|\mathbf{F}_1 + \mathbf{F}_2\|^2 = \|\mathbf{F}_1\|^2 + \|\mathbf{F}_2\|^2 + 2\|\mathbf{F}_1\|\|\mathbf{F}_2\|\cos 105° = 250^2$ and

$45° = \alpha = \sin^{-1}\left(\dfrac{\|\mathbf{F}_2\|}{250}\sin 105°\right)$, so $\dfrac{\sqrt{2}}{2} \approx \dfrac{\|\mathbf{F}_2\|}{250}0.9659, \|\mathbf{F}_2\| \approx 183.02$ lb,

$\|\mathbf{F}_1\|^2 + 2(183.02)(-0.2588)\|\mathbf{F}_1\| + (183.02)^2 = 62{,}500, \|\mathbf{F}_1\| = 224.13$ lb.

51. **(a)** $c_1\mathbf{v}_1 + c_2\mathbf{v}_2 = (2c_1 + 4c_2)\mathbf{i} + (-c_1 + 2c_2)\mathbf{j} = 4\mathbf{j}$, so $2c_1 + 4c_2 = 0$ and $-c_1 + 2c_2 = 4$ which gives $c_1 = -2, c_2 = 1$.

    (b) $c_1\mathbf{v}_1 + c_2\mathbf{v}_2 = \langle c_1 - 2c_2, -3c_1 + 6c_2\rangle = \langle 3, 5\rangle$, so $c_1 - 2c_2 = 3$ and $-3c_1 + 6c_2 = 5$ which has no solution.

53. Place $\mathbf{u}$ and $\mathbf{v}$ tip to tail so that $\mathbf{u} + \mathbf{v}$ is the vector from the initial point of $\mathbf{u}$ to the terminal point of $\mathbf{v}$. The shortest distance between two points is along the line joining these points so $\|\mathbf{u} + \mathbf{v}\| \leq \|\mathbf{u}\| + \|\mathbf{v}\|$.

55. **(d):** $\mathbf{u} + (-\mathbf{u}) = (u_1\mathbf{i} + u_2\mathbf{j}) + (-u_1\mathbf{i} - u_2\mathbf{j}) = (u_1 - u_1)\mathbf{i} + (u_1 - u_1)\mathbf{j} = \mathbf{0}$

    (g): $(k + l)\mathbf{u} = (k + l)(u_1\mathbf{i} + u_2\mathbf{j}) = ku_1\mathbf{i} + ku_2\mathbf{j} + lu_1\mathbf{i} + lu_2\mathbf{j} = k\mathbf{u} + l\mathbf{u}$

    (h): $1\mathbf{u} = 1(u_1\mathbf{i} + u_2\mathbf{j}) = 1u_1\mathbf{i} + 1u_2\mathbf{j} = u_1\mathbf{i} + u_2\mathbf{j} = \mathbf{u}$

57. Let $\mathbf{a}, \mathbf{b}, \mathbf{c}$ be vectors along the sides of the triangle and A,B the midpoints of $\mathbf{a}$ and $\mathbf{b}$, then $\mathbf{u} = \dfrac{1}{2}\mathbf{a} - \dfrac{1}{2}\mathbf{b} = \dfrac{1}{2}(\mathbf{a} - \mathbf{b}) = \dfrac{1}{2}\mathbf{c}$ so $\mathbf{u}$ is parallel to $\mathbf{c}$ and half as long.

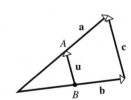

EXERCISE SET 12.3

1. **(a)** $(1)(6) + (2)(-8) = -10$; $\cos\theta = (-10)/[(\sqrt{5})(10)] = -1/\sqrt{5}$

   (b) $(-7)(0) + (-3)(1) = -3$; $\cos\theta = (-3)/[(\sqrt{58})(1)] = -3/\sqrt{58}$

   (c) $(1)(8) + (-3)(-2) + (7)(-2) = 0$; $\cos\theta = 0$

   (d) $(-3)(4) + (1)(2) + (2)(-5) = -20$; $\cos\theta = (-20)/[(\sqrt{14})(\sqrt{45})] = -20/(3\sqrt{70})$

3. **(a)** $\mathbf{u} \cdot \mathbf{v} = -34 < 0$, obtuse   **(b)** $\mathbf{u} \cdot \mathbf{v} = 6 > 0$, acute

   (c) $\mathbf{u} \cdot \mathbf{v} = -1 < 0$, obtuse   **(d)** $\mathbf{u} \cdot \mathbf{v} = 0$, orthogonal

5. Since $\mathbf{v}_0 \cdot \mathbf{v}_i = \cos\phi_i$, the answers are, in order, $\sqrt{2}/2, 0, -\sqrt{2}/2, -1, -\sqrt{2}/2, 0, \sqrt{2}/2$

7. **(a)** $\overrightarrow{AB} = \langle 1, 3, -2 \rangle$, $\overrightarrow{BC} = \langle 4, -2, -1 \rangle$, $\overrightarrow{AB} \cdot \overrightarrow{BC} = 0$ so $\overrightarrow{AB}$ and $\overrightarrow{BC}$ are orthogonal; it is a right triangle with the right angle at vertex B.

   (b) Let A, B, and C be the vertices $(-1,0)$, $(2,-1)$, and $(1,4)$ with corresponding interior angles α, β, and γ, then

   $$\cos\alpha = \frac{\overrightarrow{AB} \cdot \overrightarrow{AC}}{\|\overrightarrow{AB}\| \, \|\overrightarrow{AC}\|} = \frac{\langle 3, -1 \rangle \cdot \langle 2, 4 \rangle}{\sqrt{10}\sqrt{20}} = 1/(5\sqrt{2}), \; \alpha \approx 82°$$

   $$\cos\beta = \frac{\overrightarrow{BA} \cdot \overrightarrow{BC}}{\|\overrightarrow{BA}\| \, \|\overrightarrow{BC}\|} = \frac{\langle -3, 1 \rangle \cdot \langle -1, 5 \rangle}{\sqrt{10}\sqrt{26}} = 4/\sqrt{65}, \; \beta \approx 60°$$

   $$\cos\gamma = \frac{\overrightarrow{CA} \cdot \overrightarrow{CB}}{\|\overrightarrow{CA}\| \, \|\overrightarrow{CB}\|} = \frac{\langle -2, -4 \rangle \cdot \langle 1, -5 \rangle}{\sqrt{20}\sqrt{26}} = 9/\sqrt{130}, \; \gamma \approx 38°$$

9. **(a)** The dot product of a vector $\mathbf{u}$ and a scalar $\mathbf{v} \cdot \mathbf{w}$ is not defined.

   (b) The sum of a scalar $\mathbf{u} \cdot \mathbf{v}$ and a vector $\mathbf{w}$ is not defined.

   (c) $\mathbf{u} \cdot \mathbf{v}$ is not a vector.

   (d) The dot product of a scalar k and a vector $\mathbf{u} + \mathbf{v}$ is not defined.

11. (b): $\mathbf{u} \cdot (\mathbf{v} + \mathbf{w}) = (6\mathbf{i} - \mathbf{j} + 2\mathbf{k}) \cdot ((2\mathbf{i} + 7\mathbf{j} + 4\mathbf{k}) + (\mathbf{i} + \mathbf{j} - 3\mathbf{k})) = (6\mathbf{i} - \mathbf{j} + 2\mathbf{k}) \cdot (3\mathbf{i} + 8\mathbf{j} + \mathbf{k}) = 12$;
    $\mathbf{u} \cdot \mathbf{v} + \mathbf{u} \cdot \mathbf{w} = (6\mathbf{i} - \mathbf{j} + 2\mathbf{k}) \cdot (2\mathbf{i} + 7\mathbf{j} + 4\mathbf{k}) + (6\mathbf{i} - \mathbf{j} + 2\mathbf{k}) \cdot (\mathbf{i} + \mathbf{j} - 3\mathbf{k}) = 13 - 1 = 12$
    (c): $k(\mathbf{u} \cdot \mathbf{v}) = -5(13) = -65$; $(k\mathbf{u}) \cdot \mathbf{v} = (-30\mathbf{i} + 5\mathbf{j} - 10\mathbf{k}) \cdot (2\mathbf{i} + 7\mathbf{j} + 4\mathbf{k}) = -65$;
    $\mathbf{u} \cdot (k\mathbf{v}) = (6\mathbf{i} - \mathbf{j} + 2\mathbf{k}) \cdot (-10\mathbf{i} - 35\mathbf{j} - 20\mathbf{k}) = -65$

13. $\overrightarrow{AB} \cdot \overrightarrow{AP} = [2\mathbf{i} + \mathbf{j} + 2\mathbf{k}] \cdot [(r-1)\mathbf{i} + (r+1)\mathbf{j} + (r-3)\mathbf{k}]$

    $= 2(r-1) + (r+1) + 2(r-3) = 5r - 7 = 0, r = 7/5.$

15. **(a)** $\|\mathbf{v}\| = \sqrt{3}$ so $\cos\alpha = \cos\beta = 1/\sqrt{3}$, $\cos\gamma = -1/\sqrt{3}$, $\alpha = \beta \approx 55°$, $\gamma \approx 125°$

    (b) $\|\mathbf{v}\| = 3$ so $\cos\alpha = 2/3$, $\cos\beta = -2/3$, $\cos\gamma = 1/3$, $\alpha \approx 48°$, $\beta \approx 132°$, $\gamma \approx 71°$

17. $\cos^2\alpha + \cos^2\beta + \cos^2\gamma = \dfrac{v_1^2}{\|\mathbf{v}\|^2} + \dfrac{v_2^2}{\|\mathbf{v}\|^2} + \dfrac{v_3^2}{\|\mathbf{v}\|^2} = \left(v_1^2 + v_2^2 + v_3^2\right)/\|\mathbf{v}\|^2 = \|\mathbf{v}\|^2/\|\mathbf{v}\|^2 = 1$

19. **(a)** Let k be the length of an edge and introduce a coordinate system as shown in the figure, then $\mathbf{d} = \langle k, k, k \rangle$, $\mathbf{u} = \langle k, k, 0 \rangle$, $\cos\theta = \dfrac{\mathbf{d} \cdot \mathbf{u}}{\|\mathbf{d}\| \, \|\mathbf{u}\|} = \dfrac{2k^2}{(k\sqrt{3})(k\sqrt{2})} = 2/\sqrt{6}$
    so $\theta = \cos^{-1}(2/\sqrt{6}) \approx 35°$

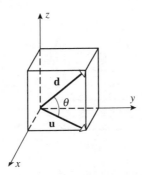

(b) $\mathbf{v} = \langle -k, 0, k \rangle, \cos\theta = \dfrac{\mathbf{d} \cdot \mathbf{v}}{\|\mathbf{d}\|\,\|\mathbf{v}\|} = 0$ so $\theta = \pi/2$ radians.

21. $\cos\alpha = \dfrac{\sqrt{3}}{2}\dfrac{1}{2} = \dfrac{\sqrt{3}}{4}$, $\cos\beta = \dfrac{\sqrt{3}}{2}\dfrac{\sqrt{3}}{2} = \dfrac{3}{4}$, $\cos\gamma = \dfrac{1}{2}$; $\alpha \approx 64°, \beta \approx 41°, \gamma = 60°$

23. Take $\mathbf{i}, \mathbf{j},$ and $\mathbf{k}$ along adjacent edges of the box, then $10\mathbf{i} + 15\mathbf{j} + 25\mathbf{k}$ is along a diagonal, and a unit vector in this direction is $\dfrac{2}{\sqrt{38}}\mathbf{i} + \dfrac{3}{\sqrt{38}}\mathbf{j} + \dfrac{5}{\sqrt{38}}\mathbf{k}$. The direction cosines are $\cos\alpha = 2/\sqrt{38}$, $\cos\beta = 3/\sqrt{38}$, and $\cos\gamma = 5/\sqrt{38}$ so $\alpha \approx 71°$, $\beta \approx 61°$, and $\gamma \approx 36°$.

25. (a) $\dfrac{\mathbf{b}}{\|\mathbf{b}\|} = \langle 1/3, 2/3, 2/3 \rangle$, so $\text{proj}_{\mathbf{b}}\mathbf{v} = \langle 2/3, 4/3, 4/3 \rangle$ and $\mathbf{v} - \text{proj}_{\mathbf{b}}\mathbf{v} = \langle 4/3, -7/3, 5/3 \rangle$

　　(b) $\dfrac{\mathbf{b}}{\|\mathbf{b}\|} = \langle 2/7, 3/7, -6/7 \rangle$, so $\text{proj}_{\mathbf{b}}\mathbf{v} = \langle -74/49, -111/49, 222/49 \rangle$
　　　 and $\mathbf{v} - \text{proj}_{\mathbf{b}}\mathbf{v} = \langle 270/49, 62/49, 121/49 \rangle$

27. (a) $\text{proj}_{\mathbf{b}}\mathbf{v} = \langle 1, 1 \rangle$, so $\mathbf{v} = \langle 1, 1 \rangle + \langle -4, 4 \rangle$
　　(b) $\text{proj}_{\mathbf{b}}\mathbf{v} = \langle 0, -8/5, 4/5 \rangle$, so $\mathbf{v} = \langle 0, -8/5, 4/5 \rangle + \langle -2, 13/5, 26/5 \rangle$
　　(c) $\mathbf{v} \cdot \mathbf{b} = 0$, hence $\text{proj}_{\mathbf{b}}\mathbf{v} = \mathbf{0}, \mathbf{v} = \mathbf{0} + \mathbf{v}$

29. $\overrightarrow{AP} = -4\mathbf{i} + 2\mathbf{k}$, $\overrightarrow{AB} = -3\mathbf{i} + 2\mathbf{j} - 4\mathbf{k}$, $\|\text{proj}_{\overrightarrow{AB}}\overrightarrow{AP}\| = |\overrightarrow{AP} \cdot \overrightarrow{AB}|/\|\overrightarrow{AB}\| = 4/\sqrt{29}$.
　　$\|\overrightarrow{AP}\| = \sqrt{20}$, $\sqrt{20 - 16/29} = \sqrt{564/29}$

31. Let x denote the magnitude of the force in the direction of $\mathbf{Q}$. Then the force $\mathbf{F}$ acting on the child is $\mathbf{F} = x\mathbf{i} - 333.2\mathbf{j}$. Let $\mathbf{e}_1 = -\langle \cos 27°, \sin 27° \rangle$ and $\mathbf{e}_2 = \langle \sin 27°, -\cos 27° \rangle$ be the unit vectors in the directions along and against the slide. Then the component of $\mathbf{F}$ in the direction of $\mathbf{e}_1$ is $\mathbf{F} \cdot \mathbf{e}_1 = -x\cos 27° + 333.2\sin 27°$ and the child is prevented from sliding down if this quantity is negative, i.e. $x > 333.2\tan 27° \approx 169.77$ N.

33. (a) Let $\mathbf{T}_A$ and $\mathbf{T}_B$ be the forces exerted on the block by cables A and B. Then $\mathbf{T}_A = a(-10\mathbf{i} + d\mathbf{j})$ and $\mathbf{T}_B = b(20\mathbf{i} + d\mathbf{j})$ for some positive a, b. Since $\mathbf{T}_A + \mathbf{T}_B - 100\mathbf{j} = \mathbf{0}$, we find $a = \dfrac{200}{3d}, b = \dfrac{100}{3d}, \mathbf{T}_A = -\dfrac{2000}{3d}\mathbf{i} + \dfrac{200}{3}\mathbf{j}$, and $\mathbf{T}_B = \dfrac{2000}{3d}\mathbf{i} + \dfrac{100}{3}\mathbf{j}$. Thus
$$\mathbf{T}_A = \dfrac{200}{3}\sqrt{1 + \dfrac{100}{d^2}}, \quad \mathbf{T}_B = \dfrac{100}{3}\sqrt{1 + \dfrac{400}{d^2}},$$
and the graphs are:

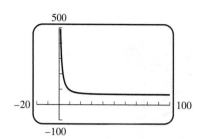
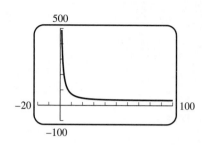

(b) An increase in d will decrease both forces.

(c) The inequality $\|\mathbf{T}_A\| \le 150$ is equivalent to $d \ge \dfrac{40}{\sqrt{65}}$, and $\|\mathbf{T}_B\| \le 150$ is equivalent to $d \ge \dfrac{40}{\sqrt{77}}$. Hence we must have $d \ge \dfrac{40}{65}$.

35. $W = \mathbf{F} \cdot (15/\sqrt{3})(\mathbf{i} + \mathbf{j} + \mathbf{k}) = -15/\sqrt{3}\ \text{N} \cdot \text{m} = -5\sqrt{3}\ \text{J}$

37. $W = \mathbf{F} \cdot 15\mathbf{i} = 15 \cdot 50 \cos 60° = 375\ \text{ft} \cdot \text{lb}.$

39. $\mathbf{u} + \mathbf{v}$ and $\mathbf{u} - \mathbf{v}$ are vectors along the diagonals,
$(\mathbf{u} + \mathbf{v}) \cdot (\mathbf{u} - \mathbf{v}) = \mathbf{u} \cdot \mathbf{u} - \mathbf{u} \cdot \mathbf{v} + \mathbf{v} \cdot \mathbf{u} - \mathbf{v} \cdot \mathbf{v} = \|\mathbf{u}\|^2 - \|\mathbf{v}\|^2$ so $(\mathbf{u} + \mathbf{v}) \cdot (\mathbf{u} - \mathbf{v}) = 0$
if and only if $\|\mathbf{u}\| = \|\mathbf{v}\|.$

41. $\|\mathbf{u} + \mathbf{v}\|^2 = (\mathbf{u} + \mathbf{v}) \cdot (\mathbf{u} + \mathbf{v}) = \|\mathbf{u}\|^2 + 2\mathbf{u} \cdot \mathbf{v} + \|\mathbf{v}\|^2$ and
$\|\mathbf{u} - \mathbf{v}\|^2 = (\mathbf{u} - \mathbf{v}) \cdot (\mathbf{u} - \mathbf{v}) = \|\mathbf{u}\|^2 - 2\mathbf{u} \cdot \mathbf{v} + \|\mathbf{v}\|^2$, add to get
$\|\mathbf{u} + \mathbf{v}\|^2 + \|\mathbf{u} - \mathbf{v}\|^2 = 2\|\mathbf{u}\|^2 + 2\|\mathbf{v}\|^2$

The sum of the squares of the lengths of the diagonals of a parallelogram is equal to twice the sum of the squares of the lengths of the sides.

43. $\mathbf{v} = c_1\mathbf{v}_1 + c_2\mathbf{v}_2 + c_3\mathbf{v}_3$ so $\mathbf{v} \cdot \mathbf{v}_i = c_i\mathbf{v}_i \cdot \mathbf{v}_i$ because $\mathbf{v}_i \cdot \mathbf{v}_j = 0$ if $i \ne j$,
thus $\mathbf{v} \cdot \mathbf{v}_i = c_i\|\mathbf{v}_i\|^2, c_i = \mathbf{v} \cdot \mathbf{v}_i/\|\mathbf{v}_i\|^2$ for $i = 1, 2, 3.$

45. **(a)** $\mathbf{u} = x\mathbf{i} + (x^2 + 1)\mathbf{j}, \mathbf{v} = x\mathbf{i} - (x + 1)\mathbf{j}, \theta = \cos^{-1}[(\mathbf{u} \cdot \mathbf{v})/(\|\mathbf{u}\|\|\mathbf{v}\|)].$

Use a CAS to solve $d\theta/dx = 0$ to find that the minimum value of θ occurs when $x \approx -3.136742$ so the minimum angle is about $40°$. NB: Since $\cos^{-1} u$ is a decreasing function of u, it suffices to maximize $(\mathbf{u} \cdot \mathbf{v})/(\|\mathbf{u}\|\|\mathbf{v}\|)$, or, what is easier, its square.

(b) Solve $\mathbf{u} \cdot \mathbf{v} = 0$ for x to get $x \approx -0.682328.$

47. Let $\mathbf{u} = \langle u_1, u_2, u_3\rangle, \mathbf{v} = \langle v_1, v_2, v_3\rangle, \mathbf{w} = \langle w_1, w_2, w_3\rangle.$ Then
$\mathbf{u} \cdot (\mathbf{v} + \mathbf{w}) = \langle u_1(v_1 + w_1), u_2(v_2 + w_2), u_3(v_3 + w_3)\rangle = \langle u_1v_1 + u_1w_1, u_2v_2 + u_2w_2, u_3v_3 + u_3w_3\rangle$
$= \langle u_1v_1, u_2v_2, u_3v_3\rangle + \langle u_1w_1, u_2w_2, u_3w_3\rangle = \mathbf{u} \cdot \mathbf{v} + \mathbf{u} \cdot \mathbf{w}$
$\mathbf{0} \cdot \mathbf{v} = 0 \cdot v_1 + 0 \cdot v_2 + 0 \cdot v_3 = 0$

EXERCISE SET 12.4

1. (a) $\mathbf{i} \times (\mathbf{i} + \mathbf{j} + \mathbf{k}) = \begin{vmatrix} \mathbf{i} & \mathbf{j} & \mathbf{k} \\ 1 & 0 & 0 \\ 1 & 1 & 1 \end{vmatrix} = -\mathbf{j} + \mathbf{k}$

   (b) $\mathbf{i} \times (\mathbf{i} + \mathbf{j} + \mathbf{k}) = (\mathbf{i} \times \mathbf{i}) + (\mathbf{i} \times \mathbf{j}) + (\mathbf{i} \times \mathbf{k}) = -\mathbf{j} + \mathbf{k}$

3. $\langle 7, 10, 9 \rangle$ 5. $\langle -4, -6, -3 \rangle$

7. (a) $\mathbf{v} \times \mathbf{w} = \langle -23, 7, -1 \rangle, \mathbf{u} \times (\mathbf{v} \times \mathbf{w}) = \langle -20, -67, -9 \rangle$

   (b) $\mathbf{u} \times \mathbf{v} = \langle -10, -14, 2 \rangle, (\mathbf{u} \times \mathbf{v}) \times \mathbf{w} = \langle -78, 52, -26 \rangle$

   (c) $(\mathbf{u} \times \mathbf{v}) \times (\mathbf{v} \times \mathbf{w}) = \langle -10, -14, 2 \rangle \times \langle -23, 7, -1 \rangle = \langle 0, -56, -392 \rangle$

   (d) $(\mathbf{v} \times \mathbf{w}) \times (\mathbf{u} \times \mathbf{v}) = \langle 0, 56, 392 \rangle$

9. $\mathbf{u} \times \mathbf{v} = (\mathbf{i} + \mathbf{j}) \times (\mathbf{i} + \mathbf{j} + \mathbf{k}) = \mathbf{k} - \mathbf{j} - \mathbf{k} + \mathbf{i} = \mathbf{i} - \mathbf{j}$, the direction cosines are $\dfrac{1}{\sqrt{2}}, -\dfrac{1}{\sqrt{2}}, 0$

11. $\mathbf{n} = \overrightarrow{AB} \times \overrightarrow{AC} = \langle 1, 1, -3 \rangle \times \langle -1, 3, -1 \rangle = \langle 8, 4, 4 \rangle$, unit vectors are $\pm \dfrac{1}{\sqrt{6}} \langle 2, 1, 1 \rangle$

13. $A = \|\mathbf{u} \times \mathbf{v}\| = \| -7\mathbf{i} - \mathbf{j} + 3\mathbf{k}\| = \sqrt{59}$

15. $A = \dfrac{1}{2} \|\overrightarrow{PQ} \times \overrightarrow{PR}\| = \dfrac{1}{2} \|\langle -1, -5, 2 \rangle \times \langle 2, 0, 3 \rangle\| = \dfrac{1}{2} \|\langle -15, 7, 10 \rangle\| = \sqrt{374}/2$

17. 80   19. -3

21. $V = |\mathbf{u} \cdot (\mathbf{v} \times \mathbf{w})| = |-16| = 16$

23. (a) $\mathbf{u} \cdot (\mathbf{v} \times \mathbf{w}) = 0$, yes   (b) $\mathbf{u} \cdot (\mathbf{v} \times \mathbf{w}) = 0$, yes   (c) $\mathbf{u} \cdot (\mathbf{v} \times \mathbf{w}) = 245$, no

25. (a) $V = |\mathbf{u} \cdot (\mathbf{v} \times \mathbf{w})| = |-9| = 9$ (b) $A = \|\mathbf{u} \times \mathbf{w}\| = \|3\mathbf{i} - 8\mathbf{j} + 7\mathbf{k}\| = \sqrt{122}$

   (c) $\mathbf{v} \times \mathbf{w} = -3\mathbf{i} - \mathbf{j} + 2\mathbf{k}$ is perpendicular to the plane determined by $\mathbf{v}$ and $\mathbf{w}$; let θ be the angle between $\mathbf{u}$ and $\mathbf{v} \times \mathbf{w}$ then

$$\cos \theta = \frac{\mathbf{u} \cdot (\mathbf{v} \times \mathbf{w})}{\|\mathbf{u}\| \, \|\mathbf{v} \times \mathbf{w}\|} = \frac{-9}{\sqrt{14}\sqrt{14}} = -9/14$$

   so the acute angle ϕ that $\mathbf{u}$ makes with the plane determined by $\mathbf{v}$ and $\mathbf{w}$ is $\phi = \theta - \pi/2 = \sin^{-1}(9/14)$.

27. (a) $\mathbf{u} = \overrightarrow{AP} = -4\mathbf{i} + 2\mathbf{k}, \mathbf{v} = \overrightarrow{AB} = -3\mathbf{i} + 2\mathbf{j} - 4\mathbf{k}, \mathbf{u} \times \mathbf{v} = -4\mathbf{i} - 22\mathbf{j} - 8\mathbf{k};$
   distance $= \|\mathbf{u} \times \mathbf{v}\|/\|\mathbf{v}\| = 2\sqrt{141/29}$

   (b) $\mathbf{u} = \overrightarrow{AP} = 2\mathbf{i} + 2\mathbf{j}, \mathbf{v} = \overrightarrow{AB} = -2\mathbf{i} + \mathbf{j}, \mathbf{u} \times \mathbf{v} = 6\mathbf{k};$ distance $= \|\mathbf{u} \times \mathbf{v}\|/\|\mathbf{v}\| = 6/\sqrt{5}$

29. $\overrightarrow{PQ} = \langle 3, -1, -3 \rangle, \overrightarrow{PR} = \langle 2, -2, 1 \rangle, \overrightarrow{PS} = \langle 4, -4, 3 \rangle,$

$V = \dfrac{1}{6} |\overrightarrow{PQ} \cdot (\overrightarrow{PR} \times \overrightarrow{PS})| = \dfrac{1}{6} |-4| = 2/3$

31. Since $\vec{AC} \cdot (\vec{AB} \times \vec{AD}) = \vec{AC} \cdot (\vec{AB} \times \vec{CD}) + \vec{AC} \cdot (\vec{AB} \times \vec{AC}) = 0 + 0 = 0$, the volume of the parallelopiped determined by $\vec{AB}, \vec{AC}$, and $\vec{AD}$ is zero, thus A, B, C, and D are coplanar (lie in the same plane). Since $\vec{AB} \times \vec{CD} \neq \mathbf{0}$, the lines are not parallel. Hence they must intersect.

33. From Theorems 12.3.3 and 12.4.5a it follows that $\sin \theta = \cos \theta$, so $\theta = \pi/4$.

35. (a) $\mathbf{F} = 10\mathbf{j}$ and $\vec{PQ} = \mathbf{i} + \mathbf{j} + \mathbf{k}$, so the vector moment of $\mathbf{F}$ about P is

$$\vec{PQ} \times \mathbf{F} = \begin{vmatrix} \mathbf{i} & \mathbf{j} & \mathbf{k} \\ 1 & 1 & 1 \\ 0 & 10 & 0 \end{vmatrix} = -10\mathbf{i} + 10\mathbf{k}, \text{ and the scalar moment is } 10\sqrt{2} \text{ lb·ft.}$$

The direction of rotation of the cube about P is counterclockwise looking along $\vec{PQ} \times \mathbf{F} = -10\mathbf{i} + 10\mathbf{k}$ toward its initial point.

(b) $\mathbf{F} = 10\mathbf{j}$ and $\vec{PQ} = \mathbf{j} + \mathbf{k}$, so the vector moment of $\mathbf{F}$ about P is

$$\vec{PQ} \times \mathbf{F} = \begin{vmatrix} \mathbf{i} & \mathbf{j} & \mathbf{k} \\ 0 & 1 & 1 \\ 0 & 10 & 0 \end{vmatrix} = -10\mathbf{i}, \text{ and the scalar moment is } 10 \text{ lb·ft. The direction of rotation}$$

of the cube about P is counterclockwise looking along $-10\mathbf{i}$ toward its initial point.

(c) $\mathbf{F} = 10\mathbf{j}$ and $\vec{PQ} = \mathbf{j}$, so the vector moment of $\mathbf{F}$ about P is

$$\vec{PQ} \times \mathbf{F} = \begin{vmatrix} \mathbf{i} & \mathbf{j} & \mathbf{k} \\ 0 & 1 & 0 \\ 0 & 10 & 0 \end{vmatrix} = \mathbf{0}, \text{ and the scalar moment is } 0 \text{ lb·ft. Since the force is parallel to}$$

the direction of motion, there is no rotation about P.

37. Take the center of the bolt as the origin of the plane. Then $\mathbf{F}$ makes an angle $72°$ with the positive x-axis, so $\mathbf{F} = 200 \cos 72°\mathbf{i} + 200 \sin 72°\mathbf{j}$ and $\vec{PQ} = 0.2\,\mathbf{i} + 0.03\,\mathbf{j}$. The scalar moment is given by

$$\left\| \begin{vmatrix} \mathbf{i} & \mathbf{j} & \mathbf{k} \\ 0.2 & 0.03 & 0 \\ 200 \cos 72° & 200 \sin 72° & 0 \end{vmatrix} \right\| = \left| 40\frac{1}{4}(\sqrt{5} - 1) - 6\frac{1}{4}\sqrt{10 + 2\sqrt{5}} \right| \approx 36.1882 \text{ N·m.}$$

39. Let $\mathbf{u} = \langle u_1, u_2, u_3 \rangle$ and $\mathbf{v} = \langle v_1, v_2, v_3 \rangle$; show that $k(\mathbf{u} \times \mathbf{v})$, $(k\mathbf{u}) \times \mathbf{v}$, and $\mathbf{u} \times (k\mathbf{v})$ are all the same; Part (e) is proved in a similar fashion.

41. $-8\mathbf{i} - 8\mathbf{k}, -8\mathbf{i} - 20\mathbf{j} + 2\mathbf{k}$

43. (a) Replace $\mathbf{u}$ with $\mathbf{a} \times \mathbf{b}$, $\mathbf{v}$ with $\mathbf{c}$, and $\mathbf{w}$ with $\mathbf{d}$ in the first formula of Exercise 41.

(b) From the second formula of Exercise 41,
$(\mathbf{a} \times \mathbf{b}) \times \mathbf{c} + (\mathbf{b} \times \mathbf{c}) \times \mathbf{a} + (\mathbf{c} \times \mathbf{a}) \times \mathbf{b}$
$= (\mathbf{c} \cdot \mathbf{a})\mathbf{b} - (\mathbf{c} \cdot \mathbf{b})\mathbf{a} + (\mathbf{a} \cdot \mathbf{b})\mathbf{c} - (\mathbf{a} \cdot \mathbf{c})\mathbf{b} + (\mathbf{b} \cdot \mathbf{c})\mathbf{a} - (\mathbf{b} \cdot \mathbf{a})\mathbf{c} = \mathbf{0}$

45. Let $\mathbf{u}$ and $\mathbf{v}$ be the vectors from a point on the curve to the points $(2, -1, 0)$ and $(3, 2, 2)$, respectively. Then $\mathbf{u} = (2 - x)\mathbf{i} + (-1 - \ln x)\mathbf{j}$ and $\mathbf{v} = (3 - x)\mathbf{i} + (2 - \ln x)\mathbf{j} + 2\mathbf{k}$. The area of the triangle is given by $A = (1/2)\|\mathbf{u} \times \mathbf{v}\|$; solve $dA/dx = 0$ for x to get $x = 2.091581$. The minimum area is 1.887850.

EXERCISE SET 12.5

In many of the Exercises in this section other answers are also possible.

1. **(a)** L_1: $P(1,0)$, $\mathbf{v} = \mathbf{j}$, $x = 1$, $y = t$
   L_2: $P(0,1)$, $\mathbf{v} = \mathbf{i}$, $x = t$, $y = 1$
   L_3: $P(0,0)$, $\mathbf{v} = \mathbf{i} + \mathbf{j}$, $x = t$, $y = t$

   (b) L_1: $P(1,1,0)$, $\mathbf{v} = \mathbf{k}$, $x = 1$, $y = 1$, $z = t$
   L_2: $P(0,1,1)$, $\mathbf{v} = \mathbf{i}$, $x = t$, $y = 1$, $z = 1$
   L_3: $P(1,0,1)$, $\mathbf{v} = \mathbf{j}$, $x = 1$, $y = t$, $z = 1$
   L_4: $P(0,0,0)$, $\mathbf{v} = \mathbf{i} + \mathbf{j} + \mathbf{k}$, $x = t$,
   $y = t$, $z = t$

3. **(a)** $\overrightarrow{P_1P_2} = \langle 2, 3 \rangle$ so $x = 3 + 2t$, $y = -2 + 3t$ for the line; for the line segment add the condition $0 \le t \le 1$.

   (b) $\overrightarrow{P_1P_2} = \langle -3, 6, 1 \rangle$ so $x = 5 - 3t$, $y = -2 + 6t$, $z = 1 + t$ for the line; for the line segment add the condition $0 \le t \le 1$.

5. **(a)** $x = 2 + t$, $y = -3 - 4t$
   (b) $x = t$, $y = -t$, $z = 1 + t$

7. **(a)** $\mathbf{r}_0 = 2\mathbf{i} - \mathbf{j}$ so $P(2, -1)$ is on the line, and $\mathbf{v} = 4\mathbf{i} - \mathbf{j}$ is parallel to the line.
   (b) At $t = 0$, $P(-1, 2, 4)$ is on the line, and $\mathbf{v} = 5\mathbf{i} + 7\mathbf{j} - 8\mathbf{k}$ is parallel to the line.

9. **(a)** $\langle x, y \rangle = \langle -3, 4 \rangle + t\langle 1, 5 \rangle$; $\mathbf{r} = -3\mathbf{i} + 4\mathbf{j} + t(\mathbf{i} + 5\mathbf{j})$
   (b) $\langle x, y, z \rangle = \langle 2, -3, 0 \rangle + t\langle -1, 5, 1 \rangle$; $\mathbf{r} = 2\mathbf{i} - 3\mathbf{j} + t(-\mathbf{i} + 5\mathbf{j} + \mathbf{k})$

11. $x = -5 + 2t$, $y = 2 - 3t$

13. $2x + 2yy' = 0$, $y' = -x/y = -(3)/(-4) = 3/4$, $\mathbf{v} = 4\mathbf{i} + 3\mathbf{j}$; $x = 3 + 4t$, $y = -4 + 3t$

15. $x = -1 + 3t$, $y = 2 - 4t$, $z = 4 + t$

17. The line is parallel to the vector $\langle 2, -1, 2 \rangle$ so $x = -2 + 2t$, $y = -t$, $z = 5 + 2t$.

19. **(a)** $y = 0$, $2 - t = 0$, $t = 2$, $x = 7$
    (b) $x = 0$, $1 + 3t = 0$, $t = -1/3$, $y = 7/3$

    (c) $y = x^2$, $2 - t = (1 + 3t)^2$, $9t^2 + 7t - 1 = 0$, $t = \dfrac{-7 \pm \sqrt{85}}{18}$, $x = \dfrac{-1 \pm \sqrt{85}}{6}$, $y = \dfrac{43 \mp \sqrt{85}}{18}$

21. **(a)** $z = 0$ when $t = 3$ so the point is $(-2, 10, 0)$
    (b) $y = 0$ when $t = -2$ so the point is $(-2, 0, -5)$
    (c) x is always -2 so the line does not intersect the yz-plane

23. $(1 + t)^2 + (3 - t)^2 = 16$, $t^2 - 2t - 3 = 0$, $(t + 1)(t - 3) = 0$; $t = -1$, 3. The points of intersection are $(0, 4, -2)$ and $(4, 0, 6)$.

25. The lines intersect if we can find values of t_1 and t_2 that satisfy the equations $2 + t_1 = 2 + t_2$, $2 + 3t_1 = 3 + 4t_2$, and $3 + t_1 = 4 + 2t_2$. Solutions of the first two of these equations are $t_1 = -1$, $t_2 = -1$ which also satisfy the third equation so the lines intersect at $(1, -1, 2)$.

27. The lines are parallel, respectively, to the vectors $\langle 7, 1, -3 \rangle$ and $\langle -1, 0, 2 \rangle$. These vectors are not parallel so the lines are not parallel. The system of equations $1 + 7t_1 = 4 - t_2$, $3 + t_1 = 6$, and $5 - 3t_1 = 7 + 2t_2$ has no solution so the lines do not intersect.

29. The lines are parallel, respectively, to the vectors $\mathbf{v}_1 = \langle -2, 1, -1 \rangle$ and $\mathbf{v}_2 = \langle -4, 2, -2 \rangle$; $\mathbf{v}_2 = 2\mathbf{v}_1$, $\mathbf{v}_1$ and $\mathbf{v}_2$ are parallel so the lines are parallel.

31. $\overrightarrow{P_1P_2} = \langle 3, -7, -7 \rangle$, $\overrightarrow{P_2P_3} = \langle -9, -7, -3 \rangle$; these vectors are not parallel so the points do not lie on the same line.

33. If t_2 gives the point $\langle -1 + 3t_2, 9 - 6t_2 \rangle$ on the second line, then $t_1 = 4 - 3t_2$ yields the point $\langle 3 - (4 - 3t_2), 1 + 2(4 - 3t_2) \rangle = \langle -1 + 3t_2, 9 - 6t_2 \rangle$ on the first line, so each point of L_2 is a point of L_1; the converse is shown with $t_2 = (4 - t_1)/3$.

35. L passes through the tips of the vectors.
$\langle x, y \rangle = \langle -1, 2 \rangle + t \langle 1, 1 \rangle$

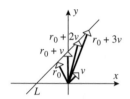

37. $\dfrac{1}{n}$ of the way from $\langle -2, 0 \rangle$ to $\langle 1, 3 \rangle$

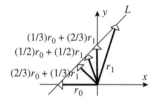

39. The line segment joining the points $(1, 0)$ and $(-3, 6)$.

41. Let the desired point be $P(x_0, y_0)$; then $\overrightarrow{P_1P} = (2/5)\,\overrightarrow{P_1P_2}$,
$\langle x_0 - 3, y_0 - 6 \rangle = (2/5)\langle 5, -10 \rangle = \langle 2, -4 \rangle$, so $x_0 = 5, y_0 = 2$.

43. $A(3, 0, 1)$ and $B(2, 1, 3)$ are on the line, and (method of Exercise 25)
$\overrightarrow{AP} = -5\mathbf{i} + \mathbf{j}$, $\overrightarrow{AB} = -\mathbf{i} + \mathbf{j} + 2\mathbf{k}$, $\|\operatorname{proj}_{\overrightarrow{AB}} \overrightarrow{AP}\| = |\overrightarrow{AP} \cdot \overrightarrow{AB}|/\|\overrightarrow{AB}\| = \sqrt{6}$ and $\|\overrightarrow{AP}\| = \sqrt{26}$,

so distance $= \sqrt{26 - 6} = 2\sqrt{5}$. Using the method of Exercise 26, distance $= \dfrac{\|\overrightarrow{AP} \times \overrightarrow{AB}\|}{\|\overrightarrow{AB}\|} = 2\sqrt{5}$.

45. The vectors $\mathbf{v}_1 = -\mathbf{i} + 2\mathbf{j} + \mathbf{k}$ and $\mathbf{v}_2 = 2\mathbf{i} - 4\mathbf{j} - 2\mathbf{k}$ are parallel to the lines, $\mathbf{v}_2 = -2\mathbf{v}_1$ so $\mathbf{v}_1$ and $\mathbf{v}_2$ are parallel. Let $t = 0$ to get the points $P(2, 0, 1)$ and $Q(1, 3, 5)$ on the first and second lines, respectively. Let $\mathbf{u} = \overrightarrow{PQ} = -\mathbf{i} + 3\mathbf{j} + 4\mathbf{k}$, $\mathbf{v} = \frac{1}{2}\mathbf{v}_2 = \mathbf{i} - 2\mathbf{j} - \mathbf{k}$; $\mathbf{u} \times \mathbf{v} = 5\mathbf{i} + 3\mathbf{j} - \mathbf{k}$; by the method of Exercise 26 of Section 12.4, distance $= \|\mathbf{u} \times \mathbf{v}\|/\|\mathbf{v}\| = \sqrt{35/6}$.

47. **(a)** The line is parallel to the vector $\langle x_1 - x_0, y_1 - y_0, z_1 - z_0 \rangle$ so
$x = x_0 + (x_1 - x_0)\,t$, $y = y_0 + (y_1 - y_0)\,t$, $z = z_0 + (z_1 - z_0)\,t$

   (b) The line is parallel to the vector $\langle a, b, c \rangle$ so $x = x_1 + at$, $y = y_1 + bt$, $z = z_1 + ct$

49. **(a)** It passes through the point $(1, -3, 5)$ and is parallel to $\mathbf{v} = 2\mathbf{i} + 4\mathbf{j} + \mathbf{k}$

   (b) $\langle x, y, z \rangle = \langle 1 + 2t, -3 + 4t, 5 + t \rangle$

51. **(a)** Let $t = 3$ and $t = -2$, respectively, in the equations for L_1 and L_2.

   (b) $\mathbf{u} = 2\mathbf{i} - \mathbf{j} - 2\mathbf{k}$ and $\mathbf{v} = \mathbf{i} + 3\mathbf{j} - \mathbf{k}$ are parallel to L_1 and L_2,
$\cos\theta = \mathbf{u} \cdot \mathbf{v}/(\|\mathbf{u}\|\,\|\mathbf{v}\|) = 1/(3\sqrt{11})$, $\theta \approx 84°$.

   (c) $\mathbf{u} \times \mathbf{v} = 7\mathbf{i} + 7\mathbf{k}$ is perpendicular to both L_1 and L_2, and hence so is $\mathbf{i} + \mathbf{k}$, thus $x = 7 + t$, $y = -1$, $z = -2 + t$.

53. $(0, 1, 2)$ is on the given line $(t = 0)$ so $\mathbf{u} = \mathbf{j} - \mathbf{k}$ is a vector from this point to the point $(0, 2, 1)$, $\mathbf{v} = 2\mathbf{i} - \mathbf{j} + \mathbf{k}$ is parallel to the given line. $\mathbf{u} \times \mathbf{v} = -2\mathbf{j} - 2\mathbf{k}$, and hence $\mathbf{w} = \mathbf{j} + \mathbf{k}$, is perpendicular to both lines so $\mathbf{v} \times \mathbf{w} = -2\mathbf{i} - 2\mathbf{j} + 2\mathbf{k}$, and hence $\mathbf{i} + \mathbf{j} - \mathbf{k}$, is parallel to the line we seek. Thus $x = t, y = 2 + t, z = 1 - t$ are parametric equations of the line.

55. **(a)** When $t = 0$ the bugs are at $(4, 1, 2)$ and $(0, 1, 1)$ so the distance between them is
$\sqrt{4^2 + 0^2 + 1^2} = \sqrt{17}$ cm.

(b)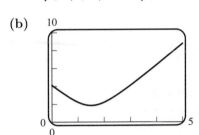

(c) The distance has a minimum value.

(d) Minimize D^2 instead of D (the distance between the bugs).
$D^2 = [t - (4 - t)]^2 + [(1 + t) - (1 + 2t)]^2 + [(1 + 2t) - (2 + t)]^2 = 6t^2 - 18t + 17$,
$d(D^2)/dt = 12t - 18 = 0$ when $t = 3/2$; the minimum
distance is $\sqrt{6(3/2)^2 - 18(3/2) + 17} = \sqrt{14}/2$ cm.

EXERCISE SET 12.6

1. $x = 3, y = 4, z = 5$

3. $(x - 2) + 4(y - 6) + 2(z - 1) = 0, x + 4y + 2z = 28$

5. $z = 0$ **7.** $\mathbf{n} = \mathbf{i} - \mathbf{j}, x - y = 0$

9. $\mathbf{n} = \mathbf{j} + \mathbf{k}, P(0, 1, 0), (y - 1) + z = 0, y + z = 1$

11. $\overrightarrow{P_1 P_2} \times \overrightarrow{P_1 P_3} = \langle 2, 1, 2 \rangle \times \langle 3, -1, -2 \rangle = \langle 0, 10, -5 \rangle$, for convenience choose $\langle 0, 2, -1 \rangle$ which is also normal to the plane. Use any of the given points to get $2y - z = 1$

13. **(a)** parallel, because $\langle 2, -8, -6 \rangle$ and $\langle -1, 4, 3 \rangle$ are parallel
(b) perpendicular, because $\langle 3, -2, 1 \rangle$ and $\langle 4, 5, -2 \rangle$ are orthogonal
(c) neither, because $\langle 1, -1, 3 \rangle$ and $\langle 2, 0, 1 \rangle$ are neither parallel nor orthogonal

15. **(a)** parallel, because $\langle 2, -1, -4 \rangle$ and $\langle 3, 2, 1 \rangle$ are orthogonal
(b) neither, because $\langle 1, 2, 3 \rangle$ and $\langle 1, -1, 2 \rangle$ are neither parallel nor orthogonal
(c) perpendicular, because $\langle 2, 1, -1 \rangle$ and $\langle 4, 2, -2 \rangle$ are parallel

17. **(a)** $3t - 2t + t - 5 = 0, t = 5/2$ so $x = y = z = 5/2$, the point of intersection is $(5/2, 5/2, 5/2)$
(b) $2(2 - t) + (3 + t) + t = 1$ has no solution so the line and plane do not intersect

19. $\mathbf{n_1} = \langle 1, 0, 0 \rangle, \mathbf{n_2} = \langle 2, -1, 1 \rangle, \mathbf{n_1} \cdot \mathbf{n_2} = 2$ so
$$\cos\theta = \frac{\mathbf{n_1} \cdot \mathbf{n_2}}{\|\mathbf{n_1}\| \, \|\mathbf{n_2}\|} = \frac{2}{\sqrt{1}\sqrt{6}} = 2/\sqrt{6}, \theta = \cos^{-1}(2/\sqrt{6}) \approx 35°$$

21. $\langle 4, -2, 7 \rangle$ is normal to the desired plane and $(0, 0, 0)$ is a point on it; $4x - 2y + 7z = 0$

23. Find two points P_1 and P_2 on the line of intersection of the given planes and then find an equation of the plane that contains P_1, P_2, and the given point $P_0(-1, 4, 2)$. Let (x_0, y_0, z_0) be on the line of intersection of the given planes; then $4x_0 - y_0 + z_0 - 2 = 0$ and $2x_0 + y_0 - 2z_0 - 3 = 0$, eliminate y_0 by addition of the equations to get $6x_0 - z_0 - 5 = 0$; if $x_0 = 0$ then $z_0 = -5$, if $x_0 = 1$ then $z_0 = 1$. Substitution of these values of x_0 and z_0 into either of the equations of the planes gives the corresponding values $y_0 = -7$ and $y_0 = 3$ so $P_1(0, -7, -5)$ and $P_2(1, 3, 1)$ are on the line of intersection of the planes. $\overrightarrow{P_0P_1} \times \overrightarrow{P_0P_2} = \langle 4, -13, 21 \rangle$ is normal to the desired plane whose equation is $4x - 13y + 21z = -14$.

25. $\mathbf{n_1} = \langle 2, 1, 1 \rangle$ and $\mathbf{n_2} = \langle 1, 2, 1 \rangle$ are normals to the given planes, $\mathbf{n_1} \times \mathbf{n_2} = \langle -1, -1, 3 \rangle$ so $\langle 1, 1, -3 \rangle$ is normal to the desired plane whose equation is $x + y - 3z = 6$.

27. $\mathbf{n_1} = \langle 2, -1, 1 \rangle$ and $\mathbf{n_2} = \langle 1, 1, -2 \rangle$ are normals to the given planes, $\mathbf{n_1} \times \mathbf{n_2} = \langle 1, 5, 3 \rangle$ is normal to the desired plane whose equation is $x + 5y + 3z = -6$.

29. The plane is the perpendicular bisector of the line segment that joins $P_1(2, -1, 1)$ and $P_2(3, 1, 5)$. The midpoint of the line segment is $(5/2, 0, 3)$ and $\overrightarrow{P_1P_2} = \langle 1, 2, 4 \rangle$ is normal to the plane so an equation is $x + 2y + 4z = 29/2$.

31. The line is parallel to the line of intersection of the planes if it is parallel to both planes. Normals to the given planes are $\mathbf{n_1} = \langle 1, -4, 2 \rangle$ and $\mathbf{n_2} = \langle 2, 3, -1 \rangle$ so $\mathbf{n_1} \times \mathbf{n_2} = \langle -2, 5, 11 \rangle$ is parallel to the line of intersection of the planes and hence parallel to the desired line whose equations are $x = 5 - 2t$, $y = 5t$, $z = -2 + 11t$.

33. $\mathbf{v_1} = \langle 1, 2, -1 \rangle$ and $\mathbf{v_2} = \langle -1, -2, 1 \rangle$ are parallel, respectively, to the given lines and to each other so the lines are parallel. Let $t = 0$ to find the points $P_1(-2, 3, 4)$ and $P_2(3, 4, 0)$ that lie, respectively, on the given lines. $\mathbf{v_1} \times \overrightarrow{P_1P_2} = \langle -7, -1, -9 \rangle$ so $\langle 7, 1, 9 \rangle$ is normal to the desired plane whose equation is $7x + y + 9z = 25$.

35. Denote the points by A, B, C, and D, respectively. The points lie in the same plane if $\overrightarrow{AB} \times \overrightarrow{AC}$ and $\overrightarrow{AB} \times \overrightarrow{AD}$ are parallel (method 1). $\overrightarrow{AB} \times \overrightarrow{AC} = \langle 0, -10, 5 \rangle$, $\overrightarrow{AB} \times \overrightarrow{AD} = \langle 0, 16, -8 \rangle$, these vectors are parallel because $\langle 0, -10, 5 \rangle = (-10/16)\langle 0, 16, -8 \rangle$. The points lie in the same plane if D lies in the plane determined by A, B, C (method 2), and since $\overrightarrow{AB} \times \overrightarrow{AC} = \langle 0, -10, 5 \rangle$, an equation of the plane is $-2y + z + 1 = 0$, $2y - z = 1$ which is satisfied by the coordinates of D.

37. Yes; if the line $L : x = a + At, y = b + Bt, z = c + Ct$ lies in a vertical plane, then the projection $L_1 : x = a + At, y = b + Bt, z = 0$ onto the x-y plane is a line (unless $A = B = 0$), and L lies in the vertical plane through L_1.
If $A = B = 0$ then $L : x = a, y = b, z = c + Ct$ lies in any vertical plane through the point (a, b, c).

39. $\mathbf{n_1} = \langle -2, 3, 7 \rangle$ and $\mathbf{n_2} = \langle 1, 2, -3 \rangle$ are normals to the planes, $\mathbf{n_1} \times \mathbf{n_2} = \langle -23, 1, -7 \rangle$ is parallel to the line of intersection. Let $z = 0$ in both equations and solve for x and y to get $x = -11/7$, $y = -12/7$ so $(-11/7, -12/7, 0)$ is on the line, a parametrization of which is $x = -11/7 - 23t$, $y = -12/7 + t$, $z = -7t$.

41. $D = |2(1) - 2(-2) + (3) - 4|/\sqrt{4 + 4 + 1} = 5/3$

43. $(0, 0, 0)$ is on the first plane so $D = |6(0) - 3(0) - 3(0) - 5|/\sqrt{36 + 9 + 9} = 5/\sqrt{54}$.

45. $(1, 3, 5)$ and $(4, 6, 7)$ are on L_1 and L_2, respectively. $\mathbf{v_1} = \langle 7, 1, -3 \rangle$ and $\mathbf{v_2} = \langle -1, 0, 2 \rangle$ are, respectively, parallel to L_1 and L_2, $\mathbf{v_1} \times \mathbf{v_2} = \langle 2, -11, 1 \rangle$ so the plane $2x - 11y + z + 51 = 0$ contains L_2 and is parallel to L_1, $D = |2(1) - 11(3) + (5) + 51|/\sqrt{4 + 121 + 1} = 25/\sqrt{126}$.

47. The distance between $(2, 1, -3)$ and the plane is $|2 - 3(1) + 2(-3) - 4|/\sqrt{1 + 9 + 4} = 11/\sqrt{14}$ which is the radius of the sphere; an equation is $(x - 2)^2 + (y - 1)^2 + (z + 3)^2 = 121/14$.

49. $\mathbf{v} = \langle 1, 2, -1 \rangle$ is parallel to the line, $\mathbf{n} = \langle 2, -2, -2 \rangle$ is normal to the plane, $\mathbf{v} \cdot \mathbf{n} = 0$ so $\mathbf{v}$ is parallel to the plane because $\mathbf{v}$ and $\mathbf{n}$ are perpendicular. $(-1, 3, 0)$ is on the line so
$$D = |2(-1) - 2(3) - 2(0) + 3|/\sqrt{4 + 4 + 4} = 5/\sqrt{12}$$

51. (a) If $\langle x_0, y_0, z_0 \rangle$ lies on the second plane, so that $ax_0 + by_0 + cz_0 + d_2 = 0$, then by Theorem 12.6.2, the distance between the planes is $D = \dfrac{|ax_0 + by_0 + cz_0 + d_1|}{\sqrt{a^2 + b^2 + c^2}} = \dfrac{|-d_2 + d_1|}{\sqrt{a^2 + b^2 + c^2}}$

(b) The distance between the planes $-2x + y + z = 0$ and $-2x + y + z + \dfrac{5}{3} = 0$ is
$$D = \frac{|0 - 5/3|}{\sqrt{4 + 1 + 1}} = \frac{5}{3\sqrt{6}}.$$

EXERCISE SET 12.7

1. (a) elliptic paraboloid, $a = 2, b = 3$
   (b) hyperbolic paraboloid, $a = 1, b = 5$
   (c) hyperboloid of one sheet, $a = b = c = 4$
   (d) circular cone, $a = b = 1$
   (e) elliptic paraboloid, $a = 2, b = 1$
   (f) hyperboloid of two sheets, $a = b = c = 1$

3. (a) $-z = x^2 + y^2$, circular paraboloid opening down the negative z-axis

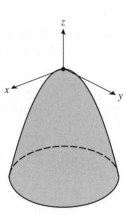

   (b) $z = x^2 + y^2$, circular paraboloid, no change
   (c) $z = x^2 + y^2$, circular paraboloid, no change

(d) $z = x^2 + y^2$, circular paraboloid, no change

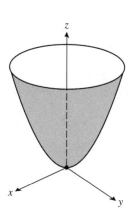

(e) $x = y^2 + z^2$, circular paraboloid opening along the positive x-axis

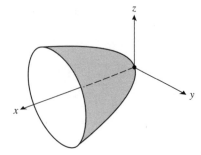

(f) $y = x^2 + z^2$, circular paraboloid opening along the positive y-axis

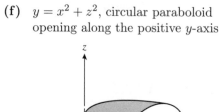

5. **(a)** hyperboloid of one sheet, axis is y-axis
   (b) hyperboloid of two sheets separated by yz-plane
   (c) elliptic paraboloid opening along the positive x-axis
   (d) elliptic cone with x-axis as axis
   (e) hyperbolic paraboloid straddling the x-axis
   (f) paraboloid opening along the negative y-axis

7. **(a)** $x = 0: \dfrac{y^2}{25} + \dfrac{z^2}{4} = 1; y = 0: \dfrac{x^2}{9} + \dfrac{z^2}{4} = 1;$

$z = 0: \dfrac{x^2}{9} + \dfrac{y^2}{25} = 1$

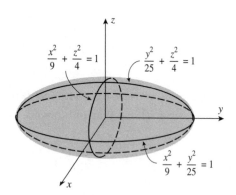

(b) $x = 0 : z = 4y^2; y = 0 : z = x^2;$

$z = 0 : x = y = 0$

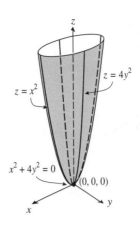

(c) $x = 0 : \dfrac{y^2}{16} - \dfrac{z^2}{4} = 1; y = 0 : \dfrac{x^2}{9} - \dfrac{z^2}{4} = 1;$

$z = 0 : \dfrac{x^2}{9} + \dfrac{y^2}{16} = 1$

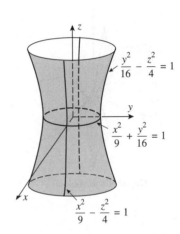

9. (a) $4x^2 + z^2 = 3$; ellipse   **(b)** $y^2 + z^2 = 3$; circle   **(c)** $y^2 + z^2 = 20$; circle

(d) $9x^2 - y^2 = 20$; hyperbola   **(e)** $z = 9x^2 + 16$; parabola   **(f)** $9x^2 + 4y^2 = 4$; ellipse

11.

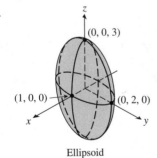

Ellipsoid

13.

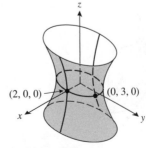

Hyperboloid
of one sheet

15.

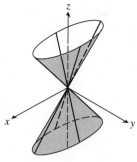

Elliptic cone

17.

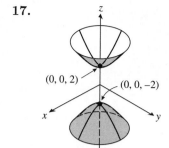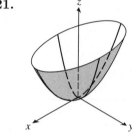

$(0, 0, 2)$ $(0, 0, -2)$

Hyperboloid
of two sheets

19.

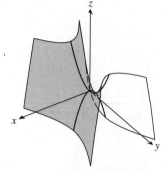

Hyperbolic paraboloid

21.

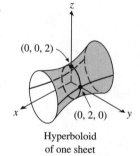

Elliptic paraboloid

23.

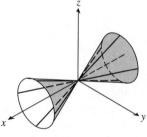

Circular cone

25.

$(0, 0, 2)$

$(0, 2, 0)$

Hyperboloid
of one sheet

27.

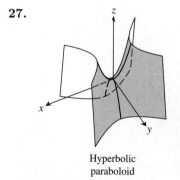

Hyperbolic
paraboloid

29.

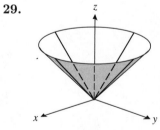

31.

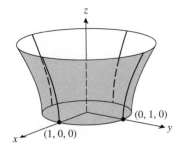

(0, 1, 0)

(1, 0, 0)

33.

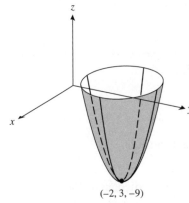

$(-2, 3, -9)$

Circular paraboloid

35.

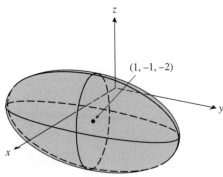

$(1, -1, -2)$

Ellipsoid

37. (a) $\dfrac{x^2}{9} + \dfrac{y^2}{4} = 1$

   (b) $6, 4$

   (c) $(\pm\sqrt{5}, 0, \sqrt{2})$

   (d) The focal axis is parallel to the x-axis.

39. (a) $\dfrac{y^2}{4} - \dfrac{x^2}{4} = 1$ (b) $(0, \pm 2, 4)$ (c) $(0, \pm 2\sqrt{2}, 4)$

   (d) The focal axis is parallel to the y-axis.

41. (a) $z + 4 = y^2$ (b) $(2, 0, -4)$ (c) $(2, 0, -15/4)$

   (d) The focal axis is parallel to the z-axis.

43. $x^2 + y^2 = 4 - x^2 - y^2, x^2 + y^2 = 2$;
   circle of radius $\sqrt{2}$ in the plane $z = 2$,
   centered at $(0, 0, 2)$

45. $y = 4(x^2 + z^2)$

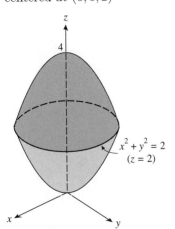

$x^2 + y^2 = 2$
$(z = 2)$

47. $|z - (-1)| = \sqrt{x^2 + y^2 + (z-1)^2}$, $z^2 + 2z + 1 = x^2 + y^2 + z^2 - 2z + 1$, $z = (x^2 + y^2)/4$; circular paraboloid

49. If $z = 0$, $\dfrac{x^2}{a^2} + \dfrac{y^2}{a^2} = 1$; if $y = 0$ then $\dfrac{x^2}{a^2} + \dfrac{z^2}{c^2} = 1$; since $c < a$ the major axis has length $2a$, the minor axis length $2c$.

51. Each slice perpendicular to the z-axis for $|z| < c$ is an ellipse whose equation is
$$\frac{x^2}{a^2} + \frac{y^2}{b^2} = \frac{c^2 - z^2}{c^2}, \text{ or } \frac{x^2}{(a^2/c^2)(c^2 - z^2)} + \frac{y^2}{(b^2/c^2)(c^2 - z^2)} = 1, \text{ the area of which is}$$
$$\pi\left(\frac{a}{c}\sqrt{c^2 - z^2}\right)\left(\frac{b}{c}\sqrt{c^2 - z^2}\right) = \pi\frac{ab}{c^2}\left(c^2 - z^2\right) \text{ so } V = 2\int_0^c \pi\frac{ab}{c^2}\left(c^2 - z^2\right)dz = \frac{4}{3}\pi abc.$$

EXERCISE SET 12.8

1. (a) $(8, \pi/6, -4)$ **(b)** $(5\sqrt{2}, 3\pi/4, 6)$ **(c)** $(2, \pi/2, 0)$ **(d)** $(8, 5\pi/3, 6)$

3. (a) $(2\sqrt{3}, 2, 3)$ **(b)** $(-4\sqrt{2}, 4\sqrt{2}, -2)$ **(c)** $(5, 0, 4)$ **(d)** $(-7, 0, -9)$

5. (a) $(2\sqrt{2}, \pi/3, 3\pi/4)$ **(b)** $(2, 7\pi/4, \pi/4)$ **(c)** $(6, \pi/2, \pi/3)$ **(d)** $(10, 5\pi/6, \pi/2)$

7. (a) $(5\sqrt{6}/4, 5\sqrt{2}/4, 5\sqrt{2}/2)$ **(b)** $(7, 0, 0)$
(c) $(0, 0, 1)$ **(d)** $(0, -2, 0)$

9. (a) $(2\sqrt{3}, \pi/6, \pi/6)$ **(b)** $(\sqrt{2}, \pi/4, 3\pi/4)$
(c) $(2, 3\pi/4, \pi/2)$ **(d)** $(4\sqrt{3}, 1, 2\pi/3)$

11. (a) $(5\sqrt{3}/2, \pi/4, -5/2)$ **(b)** $(0, 7\pi/6, -1)$
(c) $(0, 0, 3)$ **(d)** $(4, \pi/6, 0)$

15.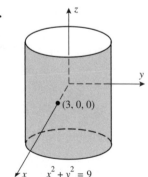

17.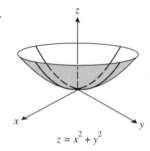

19.

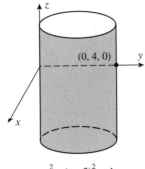

$$x^2 + (y - 2)^2 = 4$$

21.

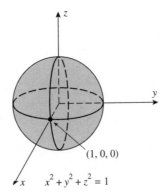

$$x^2 + y^2 + z^2 = 1$$

23.

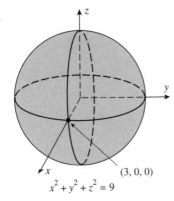

$$x^2 + y^2 + z^2 = 9$$

25.

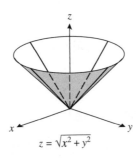

$$z = \sqrt{x^2 + y^2}$$

27.

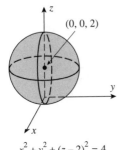

$$x^2 + y^2 + (z - 2)^2 = 4$$

29.

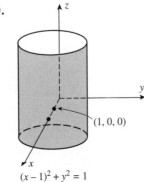

$$(x - 1)^2 + y^2 = 1$$

31. (a) $z = 3$ (b) $\rho \cos \phi = 3, \rho = 3 \sec \phi$

33. (a) $z = 3r^2$ (b) $\rho \cos \phi = 3\rho^2 \sin^2 \phi, \rho = \dfrac{1}{3} \csc \phi \cot \phi$

35. (a) $r = 2$ (b) $\rho \sin \phi = 2, \rho = 2 \csc \phi$

37. (a) $r^2 + z^2 = 9$ (b) $\rho = 3$

39. (a) $2r \cos \theta + 3r \sin \theta + 4z = 1$ (b) $2\rho \sin \phi \cos \theta + 3\rho \sin \phi \sin \theta + 4\rho \cos \phi = 1$

41. (a) $r^2 \cos^2 \theta = 16 - z^2$
      (b) $x^2 = 16 - z^2, \; x^2 + y^2 + z^2 = 16 + y^2, \; \rho^2 = 16 + \rho^2 \sin^2 \phi \sin^2 \theta, \; \rho^2 \left(1 - \sin^2 \phi \sin^2 \theta\right) = 16$

43. all points on or above the paraboloid $z = x^2 + y^2$, that are also on or below the plane $z = 4$

45. all points on or between concentric spheres of radii 1 and 3 centered at the origin

47. $\theta = \pi/6$, $\phi = \pi/6$, spherical $(4000, \pi/6, \pi/6)$, rectangular $(1000\sqrt{3}, 1000, 2000\sqrt{3})$

49. **(a)** $(10, \pi/2, 1)$ **(b)** $(0, 10, 1)$ **(c)** $(\sqrt{101}, \pi/2, \tan^{-1} 10)$

51. Using spherical coordinates: for point A, $\theta_A = 360° - 60° = 300°$, $\phi_A = 90° - 40° = 50°$; for point B, $\theta_B = 360° - 40° = 320°$, $\phi_B = 90° - 20° = 70°$. Unit vectors directed from the origin to the points A and B, respectively, are

$$\mathbf{u}_A = \sin 50° \cos 300° \mathbf{i} + \sin 50° \sin 300° \mathbf{j} + \cos 50° \mathbf{k},$$
$$\mathbf{u}_B = \sin 70° \cos 320° \mathbf{i} + \sin 70° \sin 320° \mathbf{j} + \cos 70° \mathbf{k}$$

The angle α between $\mathbf{u}_A$ and $\mathbf{u}_B$ is $\alpha = \cos^{-1}(\mathbf{u}_A \cdot \mathbf{u}_B) \approx 0.459486$ so the shortest distance is $6370\alpha \approx 2927$ km.

REVIEW EXERCISES, CHAPTER 12

3. **(b)** $x = \cos 120° = -1/2, y = \pm \sin 120° = \pm\sqrt{3}/2$

     (d) true: $\|\mathbf{u} \times \mathbf{v}\| = \|\mathbf{u}\|\|\mathbf{v}\| |\sin(\theta)| = 1$

5. $(x+3)^2 + (y-5)^2 + (z+4)^2 = r^2$,

     (a) $r^2 = 4^2 = 16$ **(b)** $r^2 = 5^2 = 25$ **(c)** $r^2 = 3^2 = 9$

7. $\overrightarrow{OS} = \overrightarrow{OP} + \overrightarrow{PS} = 3\mathbf{i} + 4\mathbf{j} + \overrightarrow{QR} = 3\mathbf{i} + 4\mathbf{j} + (4\mathbf{i} + \mathbf{j}) = 7\mathbf{i} + 5\mathbf{j}$

9. **(a)** $\mathbf{a} \cdot \mathbf{b} = 0$, $4c + 3 = 0$, $c = -3/4$

     (b) Use $\mathbf{a} \cdot \mathbf{b} = \|\mathbf{a}\| \|\mathbf{b}\| \cos\theta$ to get $4c + 3 = \sqrt{c^2 + 1}(5)\cos(\pi/4)$, $4c + 3 = 5\sqrt{c^2 + 1}/\sqrt{2}$ Square both sides and rearrange to get $7c^2 + 48c - 7 = 0$, $(7c - 1)(c + 7) = 0$ so $c = -7$ (invalid) or $c = 1/7$.

     (c) Proceed as in (b) with $\theta = \pi/6$ to get $11c^2 - 96c + 39 = 0$ and use the quadratic formula to get $c = (48 \pm 25\sqrt{3})/11$.

     (d) $\mathbf{a}$ must be a scalar multiple of $\mathbf{b}$, so $c\mathbf{i} + \mathbf{j} = k(4\mathbf{i} + 3\mathbf{j}), k = 1/3, c = 4/3$.

11. $\|\mathbf{u} - \mathbf{v}\|^2 = (\mathbf{u} - \mathbf{v}) \cdot (\mathbf{u} - \mathbf{v}) = \|\mathbf{u}\|^2 + \|\mathbf{v}\|^2 - 2\|\mathbf{u}\|\|\mathbf{v}\|\cos\theta = 2(1 - \cos\theta) = 4\sin^2(\theta/2)$, so $\|\mathbf{u} - \mathbf{v}\| = 2\sin(\theta/2)$

13. $\overrightarrow{PQ} = \langle 1, -1, 6 \rangle$, and $W = \mathbf{F} \cdot \overrightarrow{PQ} = 13$ lb·ft

15. **(a)** $\overrightarrow{AB} = -\mathbf{i} + 2\mathbf{j} + 2\mathbf{k}$, $\overrightarrow{AC} = \mathbf{i} + \mathbf{j} - \mathbf{k}$, $\overrightarrow{AB} \times \overrightarrow{AC} = -4\mathbf{i} + \mathbf{j} - 3\mathbf{k}$, area $= \frac{1}{2}\|\overrightarrow{AB} \times \overrightarrow{AC}\| = \sqrt{26}/2$

     (b) area $= \frac{1}{2}h\|\overrightarrow{AB}\| = \frac{3}{2}h = \frac{1}{2}\sqrt{26}$, $h = \sqrt{26}/3$

17. $\overrightarrow{AB} = \mathbf{i} - 2\mathbf{j} - 2\mathbf{k}$, $\overrightarrow{AC} = -2\mathbf{i} - \mathbf{j} - 2\mathbf{k}$, $\overrightarrow{AD} = \mathbf{i} + 2\mathbf{j} - 3\mathbf{k}$

     (a) From Theorem 12.4.6 and formula (9) of Section 12.4, $\begin{vmatrix} 1 & -2 & -2 \\ -2 & -1 & -2 \\ 1 & 2 & -3 \end{vmatrix} = 29$, so $V = 29$.

(b) The plane containing A, B, and C has normal $\overrightarrow{AB} \times \overrightarrow{AC} = 2\mathbf{i} + 6\mathbf{j} - 5\mathbf{k}$, so the equation of the plane is $2(x - 1) + 6(y + 1) - 5(z - 2) = 0, 2x + 6y - 5z = -14$. From Theorem 12.6.2,
$$D = \frac{|2(2) + 6(1) - 5(-1) + 14|}{\sqrt{65}} = \frac{29}{\sqrt{65}}.$$

19. $x = 4 + t, y = 1 - t, z = 2$

21. A normal to the plane is given by $\langle 1, 5, -1 \rangle$, so the equation of the plane is of the form $x + 5y - z = D$. Insert $(1, 1, 4)$ to obtain $D = 2, x + 5y - z = 2$.

23. The normals to the planes are given by $\langle a_1, b_1, c_1 \rangle$ and $\langle a_2, b_2, c_2 \rangle$, so the condition is $a_1 a_2 + b_1 b_2 + c_1 c_2 = 0$.

25. (a) $(x - 3)^2 + 4(y + 1)^2 - (z - 2)^2 = 9$, hyperboloid of one sheet

    (b) $(x + 3)^2 + (y - 2)^2 + (z + 6)^2 = 49$, sphere

    (c) $(x - 1)^2 + (y + 2)^2 - z^2 = 0$, circular cone

27. (a) $z = r^2 \cos^2 \theta - r^2 \sin^2 \theta = x^2 - y^2$ (b) $(\rho \sin \phi \cos \theta)(\rho \cos \phi) = 1, xz = 1$

29. (a) 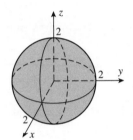   (b) 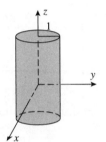   (c)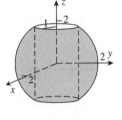

31. (a) 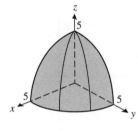   (b) 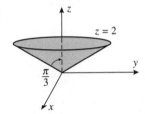   (c)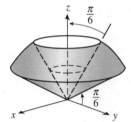

CHAPTER 13
Vector-Valued Functions

EXERCISE SET 13.1

1. $(-\infty, +\infty)$; $\mathbf{r}(\pi) = -\mathbf{i} - 3\pi\mathbf{j}$

3. $[2, +\infty)$; $\mathbf{r}(3) = -\mathbf{i} - \ln 3\mathbf{j} + \mathbf{k}$

5. $\mathbf{r} = 3\cos t\mathbf{i} + (t + \sin t)\mathbf{j}$

7. $\mathbf{r} = 2t\mathbf{i} + 2\sin 3t\mathbf{j} + 5\cos 3t\mathbf{k}$

9. $x = 3t^2$, $y = -2$

11. $x = 2t - 1$, $y = -3\sqrt{t}$, $z = \sin 3t$

13. the line in 2-space through the point $(3, 0)$ and parallel to the vector $-2\mathbf{i} + 5\mathbf{j}$

15. the line in 3-space through the point $(0, -3, 1)$ and parallel to the vector $2\mathbf{i} + 3\mathbf{k}$

17. an ellipse in the plane $z = 1$, center at $(0, 0, 1)$, major axis of length 6 parallel to y-axis, minor axis of length 4 parallel to x-axis

19. **(a)** The line is parallel to the vector $-2\mathbf{i} + 3\mathbf{j}$; the slope is $-3/2$.

(b) $y = 0$ in the xz-plane so $1 - 2t = 0$, $t = 1/2$ thus $x = 2 + 1/2 = 5/2$ and $z = 3(1/2) = 3/2$; the coordinates are $(5/2, 0, 3/2)$.

21. **(a)**

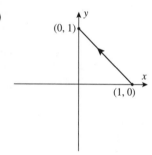

(b)

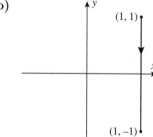

23. $\mathbf{r} = (1 - t)(3\mathbf{i} + 4\mathbf{j})$, $0 \le t \le 1$

25. $x = 2$

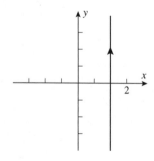

27. $(x - 1)^2 + (y - 3)^2 = 1$

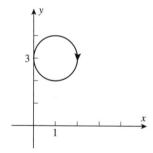

29. $x^2 - y^2 = 1$, $x \geq 1$

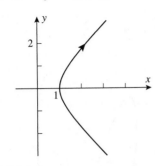

31.

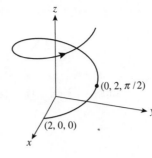

33.

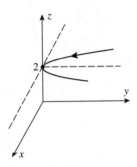

35. $x = t, y = t, z = 2t^2$

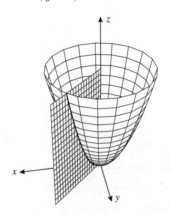

37. $\mathbf{r} = t\mathbf{i} + t^2\mathbf{j} \pm \dfrac{1}{3}\sqrt{81 - 9t^2 - t^4}\,\mathbf{k}$

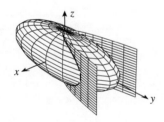

39. $x^2 + y^2 = (t\sin t)^2 + (t\cos t)^2 = t^2(\sin^2 t + \cos^2 t) = t^2 = z$

41. $x = \sin t$, $y = 2\cos t$, $z = \sqrt{3}\sin t$ so $x^2 + y^2 + z^2 = \sin^2 t + 4\cos^2 t + 3\sin^2 t = 4$ and $z = \sqrt{3}x$; it is the curve of intersection of the sphere $x^2 + y^2 + z^2 = 4$ and the plane $z = \sqrt{3}x$, which is a circle with center at $(0,0,0)$ and radius 2.

43. The helix makes one turn as t varies from 0 to 2π so $z = c(2\pi) = 3$, $c = 3/(2\pi)$.

45. $x^2 + y^2 = t^2\cos^2 t + t^2\sin^2 t = t^2$, $\sqrt{x^2 + y^2} = t = z$; a conical helix.

47. **(a)** III, since the curve is a subset of the plane $y = -x$

 (b) IV, since only x is periodic in t, and y, z increase without bound

 (c) II, since all three components are periodic in t

(d) I, since the projection onto the yz-plane is a circle and the curve increases without bound in the x-direction

49. (a) Let $x = 3\cos t$ and $y = 3\sin t$, then $z = 9\cos^2 t$.   (b)

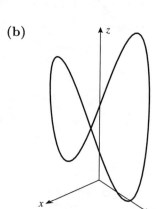

51. (a)

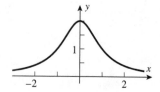

(b) In Part (a) set $x = 2t$; then $y = 2/(1 + (x/2)^2) = 8/(4 + x^2)$

EXERCISE SET 13.2

1. $\langle 1/3, 0 \rangle$

3. $2\mathbf{i} - 3\mathbf{j} + 4\mathbf{k}$

5. (a) continuous, $\lim_{t \to 0} \mathbf{r}(t) = \mathbf{0} = \mathbf{r}(0)$ (b) not continuous, $\lim_{t \to 0} \mathbf{r}(t)$ does not exist

7.

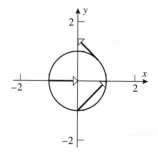

9. $\mathbf{r}'(t) = \sin t \mathbf{j}$

11. $\mathbf{r}'(t) = \langle 1, 2t \rangle$,

  $\mathbf{r}'(2) = \langle 1, 4 \rangle$,

  $\mathbf{r}(2) = \langle 2, 4 \rangle$

13. $\mathbf{r}'(t) = \sec t \tan t \mathbf{i} + \sec^2 t \mathbf{j}$,

  $\mathbf{r}'(0) = \mathbf{j}$

  $\mathbf{r}(0) = \mathbf{i}$

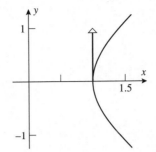

15. $\mathbf{r}'(t) = 2\cos t \mathbf{i} - 2\sin t \mathbf{k}$,

  $\mathbf{r}'(\pi/2) = -2\mathbf{k}$,

  $\mathbf{r}(\pi/2) = 2\mathbf{i} + \mathbf{j}$

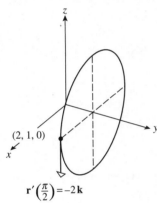

17.

19. $\mathbf{r}'(t) = 2t\mathbf{i} - \dfrac{1}{t}\mathbf{j}$, $\mathbf{r}'(1) = 2\mathbf{i} - \mathbf{j}$, $\mathbf{r}(1) = \mathbf{i} + 2\mathbf{j}$; $x = 1 + 2t$, $y = 2 - t$

21. $\mathbf{r}'(t) = -2\pi \sin \pi t \mathbf{i} + 2\pi \cos \pi t \mathbf{j} + 3\mathbf{k}$, $\mathbf{r}'(1/3) = -\sqrt{3}\,\pi \mathbf{i} + \pi \mathbf{j} + 3\mathbf{k}$,

  $\mathbf{r}(1/3) = \mathbf{i} + \sqrt{3}\,\mathbf{j} + \mathbf{k}$; $x = 1 - \sqrt{3}\,\pi t$, $y = \sqrt{3} + \pi t$, $z = 1 + 3t$

23. $\mathbf{r}'(t) = 2\mathbf{i} + \dfrac{3}{2\sqrt{3t + 4}}\mathbf{j}$, $t = 0$ at P_0 so $\mathbf{r}'(0) = 2\mathbf{i} + \dfrac{3}{4}\mathbf{j}$,

  $\mathbf{r}(0) = -\mathbf{i} + 2\mathbf{j}$; $\mathbf{r} = (-\mathbf{i} + 2\mathbf{j}) + t\left(2\mathbf{i} + \dfrac{3}{4}\mathbf{j}\right)$

25. $\mathbf{r}'(t) = 2t\mathbf{i} + \dfrac{1}{(t + 1)^2}\mathbf{j} - 2t\mathbf{k}$, $t = -2$ at P_0 so $\mathbf{r}'(-2) = -4\mathbf{i} + \mathbf{j} + 4\mathbf{k}$,

  $\mathbf{r}(-2) = 4\mathbf{i} + \mathbf{j}$; $\mathbf{r} = (4\mathbf{i} + \mathbf{j}) + t(-4\mathbf{i} + \mathbf{j} + 4\mathbf{k})$

27.   **(a)** $\lim\limits_{t \to 0}(\mathbf{r}(t) - \mathbf{r}'(t)) = \mathbf{i} - \mathbf{j} + \mathbf{k}$

  (b) $\lim\limits_{t \to 0}(\mathbf{r}(t) \times \mathbf{r}'(t)) = \lim\limits_{t \to 0}(-\cos t \mathbf{i} - \sin t \mathbf{j} + \mathbf{k}) = -\mathbf{i} + \mathbf{k}$

  (c) $\lim\limits_{t \to 0}(\mathbf{r}(t) \cdot \mathbf{r}'(t)) = 0$

29. (a) $\mathbf{r}_1' = 2\mathbf{i} + 6t\mathbf{j} + 3t^2\mathbf{k}$, $\mathbf{r}_2' = 4t^3\mathbf{k}$, $\mathbf{r}_1 \cdot \mathbf{r}_2 = t^7$; $\dfrac{d}{dt}(\mathbf{r}_1 \cdot \mathbf{r}_2) = 7t^6 = \mathbf{r}_1 \cdot \mathbf{r}_2' + \mathbf{r}_1' \cdot \mathbf{r}_2$

(b) $\mathbf{r}_1 \times \mathbf{r}_2 = 3t^6\mathbf{i} - 2t^5\mathbf{j}$, $\dfrac{d}{dt}(\mathbf{r}_1 \times \mathbf{r}_2) = 18t^5\mathbf{i} - 10t^4\mathbf{j} = \mathbf{r}_1 \times \mathbf{r}_2' + \mathbf{r}_1' \times \mathbf{r}_2$

31. $3t\mathbf{i} + 2t^2\mathbf{j} + \mathbf{C}$

33. $(-t\cos t + \sin t)\mathbf{i} + t\mathbf{j} + \mathbf{C}$

35. $(t^3/3)\mathbf{i} - t^2\mathbf{j} + \ln|t|\mathbf{k} + \mathbf{C}$

37. $\left\langle \dfrac{1}{2}\sin 2t, -\dfrac{1}{2}\cos 2t \right\rangle \Big]_0^{\pi/2} = \langle 0, 1 \rangle$

39. $\displaystyle\int_0^2 \sqrt{t^2 + t^4}\,dt = \int_0^2 t(1 + t^2)^{1/2}\,dt = \dfrac{1}{3}\left(1 + t^2\right)^{3/2}\bigg]_0^2 = (5\sqrt{5} - 1)/3$

41. $\left(\dfrac{2}{3}t^{3/2}\mathbf{i} + 2t^{1/2}\mathbf{j}\right)\bigg]_1^9 = \dfrac{52}{3}\mathbf{i} + 4\mathbf{j}$

43. $\mathbf{y}(t) = \displaystyle\int \mathbf{y}'(t)\,dt = t^2\mathbf{i} + t^3\mathbf{j} + \mathbf{C}, \mathbf{y}(0) = \mathbf{C} = \mathbf{i} - \mathbf{j}, \mathbf{y}(t) = (t^2 + 1)\mathbf{i} + (t^3 - 1)\mathbf{j}$

45. $\mathbf{y}'(t) = \displaystyle\int \mathbf{y}''(t)\,dt = t\mathbf{i} + e^t\mathbf{j} + \mathbf{C}_1, \mathbf{y}'(0) = \mathbf{j} + \mathbf{C}_1 = \mathbf{j}$ so $\mathbf{C}_1 = \mathbf{0}$ and $\mathbf{y}'(t) = t\mathbf{i} + e^t\mathbf{j}$.

$\mathbf{y}(t) = \displaystyle\int \mathbf{y}'(t)\,dt = \dfrac{1}{2}t^2\mathbf{i} + e^t\mathbf{j} + \mathbf{C}_2, \mathbf{y}(0) = \mathbf{j} + \mathbf{C}_2 = 2\mathbf{i}$ so $\mathbf{C}_2 = 2\mathbf{i} - \mathbf{j}$ and

$\mathbf{y}(t) = \left(\dfrac{1}{2}t^2 + 2\right)\mathbf{i} + (e^t - 1)\mathbf{j}$

47. $\mathbf{r}'(t) = -4\sin t\,\mathbf{i} + 3\cos t\,\mathbf{j}$, $\mathbf{r}(t) \cdot \mathbf{r}'(t) = -7\cos t \sin t$, so $\mathbf{r}$ and $\mathbf{r}'$ are perpendicular for $t = 0, \pi/2, \pi, 3\pi/2, 2\pi$. Since

$\|\mathbf{r}(t)\| = \sqrt{16\cos^2 t + 9\sin^2 t}$, $\|\mathbf{r}'(t)\| = \sqrt{16\sin^2 t + 9\cos^2 t}$,

$\|\mathbf{r}\|\|\mathbf{r}'\| = \sqrt{144 + 337\sin^2 t\cos^2 t}$, $\theta = \cos^{-1}\left[\dfrac{-7\sin t\cos t}{\sqrt{144 + 337\sin^2 t\cos^2 t}}\right]$, with the graph

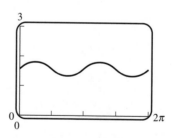

From the graph it appears that θ is bounded away from 0 and π, meaning that $\mathbf{r}$ and $\mathbf{r}'$ are never parallel. We can check this by considering them as vectors in 3-space, and then $\mathbf{r} \times \mathbf{r}' = 12\,\mathbf{k} \neq \mathbf{0}$, so they are never parallel.

49. (a) $2t - t^2 - 3t = -2$, $t^2 + t - 2 = 0$, $(t + 2)(t - 1) = 0$ so $t = -2, 1$. The points of intersection are $(-2, 4, 6)$ and $(1, 1, -3)$.

(b) $r' = i + 2tj - 3k$; $r'(-2) = i - 4j - 3k$, $r'(1) = i + 2j - 3k$, and $n = 2i - j + k$ is normal to
the plane. Let θ be the acute angle, then
for $t = -2$: $\cos\theta = |n \cdot r'|/(\|n\| \|r'\|) = 3/\sqrt{156}$, $\theta \approx 76°$;
for $t = 1$: $\cos\theta = |n \cdot r'|/(\|n\| \|r'\|) = 3/\sqrt{84}$, $\theta \approx 71°$.

51. $r_1(1) = r_2(2) = i + j + 3k$ so the graphs intersect at P; $r_1'(t) = 2ti + j + 9t^2k$ and

$r_2'(t) = i + \frac{1}{2}tj - k$ so $r_1'(1) = 2i + j + 9k$ and $r_2'(2) = i + j - k$ are tangent to the graphs at P,

thus $\cos\theta = \dfrac{r_1'(1) \cdot r_2'(2)}{\|r_1'(1)\| \|r_2'(2)\|} = -\dfrac{6}{\sqrt{86}\sqrt{3}}$, $\theta = \cos^{-1}(6/\sqrt{258}) \approx 68°$.

53. $\dfrac{d}{dt}[r(t) \times r'(t)] = r(t) \times r''(t) + r'(t) \times r'(t) = r(t) \times r''(t) + 0 = r(t) \times r''(t)$

55. In Exercise 54, write each scalar triple product as a determinant.

57. Let $r_1(t) = x_1(t)i + y_1(t)j + z_1(t)k$ and $r_2(t) = x_2(t)i + y_2(t)j + z_2(t)k$, in both (6) and (7); show
that the left and right members of the equalities are the same.

EXERCISE SET 13.3

1. $r'(t) = 3t^2i + (6t - 2)j + 2tk$; smooth

3. $r'(t) = (1 - t)e^{-t}i + (2t - 2)j - \pi\sin(\pi t)k$; not smooth, $r'(1) = 0$

5. $(dx/dt)^2 + (dy/dt)^2 + (dz/dt)^2 = (-3\cos^2 t \sin t)^2 + (3\sin^2 t \cos t)^2 + 0^2 = 9\sin^2 t \cos^2 t$,

$L = \displaystyle\int_0^{\pi/2} 3\sin t \cos t \, dt = 3/2$

7. $r'(t) = \langle e^t, -e^{-t}, \sqrt{2}\rangle$, $\|r'(t)\| = e^t + e^{-t}$, $L = \displaystyle\int_0^1 (e^t + e^{-t})dt = e - e^{-1}$

9. $r'(t) = 3t^2i + j + \sqrt{6}\,tk$, $\|r'(t)\| = 3t^2 + 1$, $L = \displaystyle\int_1^3 (3t^2 + 1)dt = 28$

11. $r'(t) = -3\sin ti + 3\cos tj + k$, $\|r'(t)\| = \sqrt{10}$, $L = \displaystyle\int_0^{2\pi} \sqrt{10}\, dt = 2\pi\sqrt{10}$

13. $(dr/dt)(dt/d\tau) = (i + 2tj)(4) = 4i + 8tj = 4i + 8(4\tau + 1)j$;
$r(\tau) = (4\tau + 1)i + (4\tau + 1)^2 j$, $r'(\tau) = 4i + 2(4)(4\tau + 1)j$

15. $(dr/dt)(dt/d\tau) = (e^ti - 4e^{-t}j)(2\tau) = 2\tau e^{\tau^2}i - 8\tau e^{-\tau^2}j$;
$r(\tau) = e^{\tau^2}i + 4e^{-\tau^2}j$, $r'(\tau) = 2\tau e^{\tau^2}i - 4(2)\tau e^{-\tau^2}j$

17. **(a)** The tangent vector reverses direction at the four cusps.
(b) $r'(t) = -3\cos^2 t \sin ti + 3\sin^2 t \cos tj = 0$ when $t = 0, \pi/2, \pi, 3\pi/2, 2\pi$.

19. (a) $\|\mathbf{r}'(t)\| = \sqrt{2}, s = \int_0^t \sqrt{2}\, dt = \sqrt{2}t; \mathbf{r} = \dfrac{s}{\sqrt{2}}\mathbf{i} + \dfrac{s}{\sqrt{2}}\mathbf{j}, x = \dfrac{s}{\sqrt{2}}, y = \dfrac{s}{\sqrt{2}}$

(b) Similar to Part (a), $x = y = z = \dfrac{s}{\sqrt{3}}$

21. (a) $\mathbf{r}(t) = \langle 1, 3, 4 \rangle$ when $t = 0$,

so $s = \int_0^t \sqrt{1 + 4 + 4}\, du = 3t, x = 1 + s/3, y = 3 - 2s/3, z = 4 + 2s/3$

(b) $\mathbf{r}\Big]_{s=25} = \langle 28/3, -41/3, 62/3 \rangle$

23. $x = 3 + \cos t, y = 2 + \sin t, (dx/dt)^2 + (dy/dt)^2 = 1,$

$s = \int_0^t du = t$ so $t = s, x = 3 + \cos s, y = 2 + \sin s$ for $0 \leq s \leq 2\pi$.

25. $x = t^3/3, y = t^2/2, (dx/dt)^2 + (dy/dt)^2 = t^2(t^2 + 1),$

$s = \int_0^t u(u^2 + 1)^{1/2} du = \dfrac{1}{3}[(t^2 + 1)^{3/2} - 1]$ so $t = [(3s + 1)^{2/3} - 1]^{1/2},$

$x = \dfrac{1}{3}[(3s + 1)^{2/3} - 1]^{3/2}, y = \dfrac{1}{2}[(3s + 1)^{2/3} - 1]$ for $s \geq 0$

27. $x = e^t \cos t, y = e^t \sin t, (dx/dt)^2 + (dy/dt)^2 = 2e^{2t}, s = \int_0^t \sqrt{2}\, e^u\, du = \sqrt{2}(e^t - 1)$ so

$t = \ln(s/\sqrt{2} + 1), x = (s/\sqrt{2} + 1)\cos[\ln(s/\sqrt{2} + 1)], y = (s/\sqrt{2} + 1)\sin[\ln(s/\sqrt{2} + 1)]$
for $0 \leq s \leq \sqrt{2}(e^{\pi/2} - 1)$

29. $dx/dt = -a\sin t, dy/dt = a\cos t, dz/dt = c,$

$s(t_0) = L = \int_0^{t_0} \sqrt{a^2 \sin^2 t + a^2 \cos^2 t + c^2}\, dt = \int_0^{t_0} \sqrt{a^2 + c^2}\, dt = t_0 \sqrt{a^2 + c^2}$

31. $x = at - a\sin t, y = a - a\cos t, (dx/dt)^2 + (dy/dt)^2 = 4a^2 \sin^2(t/2),$

$s = \int_0^t 2a\sin(u/2)du = 4a[1 - \cos(t/2)]$ so $\cos(t/2) = 1 - s/(4a), t = 2\cos^{-1}[1 - s/(4a)],$

$\cos t = 2\cos^2(t/2) - 1 = 2[1 - s/(4a)]^2 - 1,$

$\sin t = 2\sin(t/2)\cos(t/2) = 2(1 - [1 - s/(4a)]^2)^{1/2}(2[1 - s/(4a)]^2 - 1),$

$x = 2a\cos^{-1}[1 - s/(4a)] - 2a(1 - [1 - s/(4a)]^2)^{1/2}(2[1 - s/(4a)]^2 - 1),$

$y = \dfrac{s(8a - s)}{8a}$ for $0 \leq s \leq 8a$

33. (a) $(dr/dt)^2 + r^2(d\theta/dt)^2 + (dz/dt)^2 = 9e^{4t}, L = \int_0^{\ln 2} 3e^{2t}dt = \dfrac{3}{2}e^{2t}\Big]_0^{\ln 2} = 9/2$

(b) $(dr/dt)^2 + r^2(d\theta/dt)^2 + (dz/dt)^2 = 5t^2 + t^4 = t^2(5 + t^2),$

$L = \int_1^2 t(5 + t^2)^{1/2}dt = 9 - 2\sqrt{6}$

35. (a) $(d\rho/dt)^2 + \rho^2 \sin^2 \phi (d\theta/dt)^2 + \rho^2 (d\phi/dt)^2 = 3e^{-2t}$, $L = \int_0^2 \sqrt{3}e^{-t}dt = \sqrt{3}(1 - e^{-2})$

   (b) $(d\rho/dt)^2 + \rho^2 \sin^2 \phi (d\theta/dt)^2 + \rho^2 (d\phi/dt)^2 = 5$, $L = \int_1^5 \sqrt{5}dt = 4\sqrt{5}$

37. (a) $g(\tau) = \pi\tau$ **(b)** $g(\tau) = \pi(1 - \tau)$

39. Represent the helix by $x = a\cos t$, $y = a\sin t$, $z = ct$ with $a = 6.25$ and $c = 10/\pi$, so that the radius of the helix is the distance from the axis of the cylinder to the center of the copper cable, and the helix makes one turn in a distance of 20 in. $(t = 2\pi)$. From Exercise 29 the length of the helix is $2\pi\sqrt{6.25^2 + (10/\pi)^2} \approx 44$ in.

41. $\mathbf{r}'(t) = (1/t)\mathbf{i} + 2\mathbf{j} + 2t\mathbf{k}$

   (a) $\|\mathbf{r}'(t)\| = \sqrt{1/t^2 + 4 + 4t^2} = \sqrt{(2t + 1/t)^2} = 2t + 1/t$

   (b) $\dfrac{ds}{dt} = 2t + 1/t$ **(c)** $\int_1^3 (2t + 1/t)dt = 8 + \ln 3$

43. Let $\mathbf{r}(t) = x(t)\mathbf{i} + y(t)\mathbf{j}$ and use the chain rule.

EXERCISE SET 13.4

1. (a)

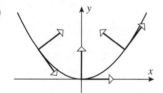

   (b)

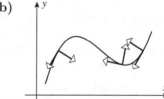

3. From the marginal note, the line is parametrized by normalizing $\mathbf{v}$, but $\mathbf{T}(t_0) = \mathbf{v}/\|\mathbf{v}\|$, so $\mathbf{r} = \mathbf{r}(t_0) + t\mathbf{v}$ becomes $\mathbf{r} = \mathbf{r}(t_0) + s\mathbf{T}(t_0)$.

5. $\mathbf{r}'(t) = 2t\mathbf{i} + \mathbf{j}$, $\|\mathbf{r}'(t)\| = \sqrt{4t^2 + 1}$, $\mathbf{T}(t) = (4t^2 + 1)^{-1/2}(2t\mathbf{i} + \mathbf{j})$,
   $\mathbf{T}'(t) = (4t^2 + 1)^{-1/2}(2\mathbf{i}) - 4t(4t^2 + 1)^{-3/2}(2t\mathbf{i} + \mathbf{j})$;
   $\mathbf{T}(1) = \dfrac{2}{\sqrt{5}}\mathbf{i} + \dfrac{1}{\sqrt{5}}\mathbf{j}$, $\mathbf{T}'(1) = \dfrac{2}{5\sqrt{5}}(\mathbf{i} - 2\mathbf{j})$, $\mathbf{N}(1) = \dfrac{1}{\sqrt{5}}\mathbf{i} - \dfrac{2}{\sqrt{5}}\mathbf{j}$.

7. $\mathbf{r}'(t) = -5\sin t\mathbf{i} + 5\cos t\mathbf{j}$, $\|\mathbf{r}'(t)\| = 5$, $\mathbf{T}(t) = -\sin t\mathbf{i} + \cos t\mathbf{j}$, $\mathbf{T}'(t) = -\cos t\mathbf{i} - \sin t\mathbf{j}$;
   $\mathbf{T}(\pi/3) = -\dfrac{\sqrt{3}}{2}\mathbf{i} + \dfrac{1}{2}\mathbf{j}$, $\mathbf{T}'(\pi/3) = -\dfrac{1}{2}\mathbf{i} - \dfrac{\sqrt{3}}{2}\mathbf{j}$, $\mathbf{N}(\pi/3) = -\dfrac{1}{2}\mathbf{i} - \dfrac{\sqrt{3}}{2}\mathbf{j}$

9. $\mathbf{r}'(t) = -4\sin t\mathbf{i} + 4\cos t\mathbf{j} + \mathbf{k}$, $\mathbf{T}(t) = \dfrac{1}{\sqrt{17}}(-4\sin t\mathbf{i} + 4\cos t\mathbf{j} + \mathbf{k})$,
   $\mathbf{T}'(t) = \dfrac{1}{\sqrt{17}}(-4\cos t\mathbf{i} - 4\sin t\mathbf{j})$, $\mathbf{T}(\pi/2) = -\dfrac{4}{\sqrt{17}}\mathbf{i} + \dfrac{1}{\sqrt{17}}\mathbf{k}$
   $\mathbf{T}'(\pi/2) = -\dfrac{4}{\sqrt{17}}\mathbf{j}$, $\mathbf{N}(\pi/2) = -\mathbf{j}$

11. $\mathbf{r}'(t) = e^t[(\cos t - \sin t)\mathbf{i} + (\cos t + \sin t)\mathbf{j} + \mathbf{k}]$, $\mathbf{T}(t) = \dfrac{1}{\sqrt{3}}[(\cos t - \sin t)\mathbf{i} + (\cos t + \sin t)\mathbf{j} + \mathbf{k}]$,

$\mathbf{T}'(t) = \dfrac{1}{\sqrt{3}}[(-\sin t - \cos t)\mathbf{i} + (-\sin t + \cos t)\mathbf{j}]$,

$\mathbf{T}(0) = \dfrac{1}{\sqrt{3}}\mathbf{i} + \dfrac{1}{\sqrt{3}}\mathbf{j} + \dfrac{1}{\sqrt{3}}\mathbf{k}$, $\mathbf{T}'(0) = \dfrac{1}{\sqrt{3}}(-\mathbf{i} + \mathbf{j})$, $\mathbf{N}(0) = -\dfrac{1}{\sqrt{2}}\mathbf{i} + \dfrac{1}{\sqrt{2}}\mathbf{j}$

13. $\mathbf{r}'(t) = \cos t\,\mathbf{i} - \sin t\,\mathbf{j} + t\,\mathbf{k}$, $\mathbf{r}'(0) = \mathbf{i}$, $\mathbf{r}(0) = \mathbf{j}$, $\mathbf{T}(0) = \mathbf{i}$, so the tangent line has the parametrization $x = s, y = 1$.

15. $\mathbf{T} = \dfrac{3}{5}\cos t\,\mathbf{i} - \dfrac{3}{5}\sin t\,\mathbf{j} + \dfrac{4}{5}\mathbf{k}$, $\mathbf{N} = -\sin t\,\mathbf{i} - \cos t\,\mathbf{j}$, $\mathbf{B} = \mathbf{T} \times \mathbf{N} = \dfrac{4}{5}\cos t\,\mathbf{i} - \dfrac{4}{5}\sin t\,\mathbf{j} - \dfrac{3}{5}\mathbf{k}$. Check:

$\mathbf{r}' = 3\cos t\,\mathbf{i} - 3\sin t\,\mathbf{j} + 4\,\mathbf{k}$, $\mathbf{r}'' = -3\sin t\,\mathbf{i} - 3\cos t\,\mathbf{j}$, $\mathbf{r}' \times \mathbf{r}'' = 12\cos t\,\mathbf{i} - 12\sin t\,\mathbf{j} - 9\,\mathbf{k}$,

$\|\mathbf{r}' \times \mathbf{r}''\| = 15$, $(\mathbf{r}' \times \mathbf{r}'')/\|\mathbf{r}' \times \mathbf{r}''\| = \dfrac{4}{5}\cos t\,\mathbf{i} - \dfrac{4}{5}\sin t\,\mathbf{j} - \dfrac{3}{5}\mathbf{k} = \mathbf{B}$.

17. $\mathbf{r}'(t) = t\sin t\,\mathbf{i} + t\cos t\,\mathbf{j}$, $\|\mathbf{r}'\| = t$, $\mathbf{T} = \sin t\,\mathbf{i} + \cos t\,\mathbf{j}$, $\mathbf{N} = \cos t\,\mathbf{i} - \sin t\,\mathbf{j}$, $\mathbf{B} = \mathbf{T} \times \mathbf{N} = -\mathbf{k}$. Check:

$\mathbf{r}' = t\sin t\,\mathbf{i} + t\cos t\,\mathbf{j}$, $\mathbf{r}'' = (\sin t + t\cos t)\,\mathbf{i} + (\cos t - t\sin t)\,\mathbf{j}$, $\mathbf{r}' \times \mathbf{r}'' = -2e^{2t}\,\mathbf{k}$,

$\|\mathbf{r}' \times \mathbf{r}''\| = 2e^{2t}$, $(\mathbf{r}' \times \mathbf{r}'')/\|\mathbf{r}' \times \mathbf{r}''\| = -\mathbf{k} = \mathbf{B}$.

19. $\mathbf{r}(\pi/4) = \dfrac{\sqrt{2}}{2}\mathbf{i} + \dfrac{\sqrt{2}}{2}\mathbf{j} + \mathbf{k}$, $\mathbf{T} = -\sin t\,\mathbf{i} + \cos t\,\mathbf{j} = \dfrac{\sqrt{2}}{2}(-\mathbf{i} + \mathbf{j})$, $\mathbf{N} = -(\cos t\,\mathbf{i} + \sin t\,\mathbf{j}) = -\dfrac{\sqrt{2}}{2}(\mathbf{i} + \mathbf{j})$,

$\mathbf{B} = \mathbf{k}$; the rectifying, osculating, and normal planes are given (respectively) by $x + y = \sqrt{2}$, $z = 1, -x + y = 0$.

21. (a) By formulae (1) and (11), $\mathbf{N}(t) = \mathbf{B}(t) \times \mathbf{T}(t) = \dfrac{\mathbf{r}'(t) \times \mathbf{r}''(t)}{\|\mathbf{r}'(t) \times \mathbf{r}''(t)\|} \times \dfrac{\mathbf{r}'(t)}{\|\mathbf{r}'(t)\|}$.

(b) Since $\mathbf{r}'$ is perpendicular to $\mathbf{r}' \times \mathbf{r}''$ it follows from Lagrange's Identity (Exercise 34 of Section 12.4) that $\|(\mathbf{r}'(t) \times \mathbf{r}''(t)) \times \mathbf{r}'(t)\| = \|\mathbf{r}'(t) \times \mathbf{r}''(t)\|\|\mathbf{r}'(t)\|$, and the result follows.

(c) From Exercise 41 of Section 12.4, $(\mathbf{r}'(t) \times \mathbf{r}''(t)) \times \mathbf{r}'(t) = \|\mathbf{r}'(t)\|^2\mathbf{r}''(t) - (\mathbf{r}'(t) \cdot \mathbf{r}''(t))\mathbf{r}'(t) = \mathbf{u}(t)$, so $\mathbf{N}(t) = \mathbf{u}(t)/\|\mathbf{u}(t)\|$

23. $\mathbf{r}'(t) = \cos t\,\mathbf{i} - \sin t\,\mathbf{j} + \mathbf{k}$, $\mathbf{r}''(t) = -\sin t\,\mathbf{i} - \cos t\,\mathbf{j}$, $\mathbf{u} = -2(\sin t\,\mathbf{i} + \cos t\,\mathbf{j})$, $\|\mathbf{u}\| = 2$, $\mathbf{N} = -\sin t\,\mathbf{i} - \cos t\,\mathbf{j}$

EXERCISE SET 13.5

1. $\kappa \approx \dfrac{1}{0.5} = 2$

3. (a) At $x = 0$ the curvature of I has a large value, yet the value of II there is zero, so II is not the curvature of I; hence I is the curvature of II.

(b) I has points of inflection where the curvature is zero, but II is not zero there, and hence is not the curvature of I; so I is the curvature of II.

5. $\mathbf{r}'(t) = 2t\mathbf{i} + 3t^2\mathbf{j}$, $\mathbf{r}''(t) = 2\mathbf{i} + 6t\mathbf{j}$, $\kappa = \|\mathbf{r}'(t) \times \mathbf{r}''(t)\|/\|\mathbf{r}'(t)\|^3 = \dfrac{6}{t(4 + 9t^2)^{3/2}}$

7. $\mathbf{r}'(t) = 3e^{3t}\mathbf{i} - e^{-t}\mathbf{j}$, $\mathbf{r}''(t) = 9e^{3t}\mathbf{i} + e^{-t}\mathbf{j}$, $\kappa = \|\mathbf{r}'(t) \times \mathbf{r}''(t)\|/\|\mathbf{r}'(t)\|^3 = \dfrac{12e^{2t}}{(9e^{6t} + e^{-2t})^{3/2}}$

9. $\mathbf{r}'(t) = -4\sin t\mathbf{i} + 4\cos t\mathbf{j} + \mathbf{k}$, $\mathbf{r}''(t) = -4\cos t\mathbf{i} - 4\sin t\mathbf{j}$,
   $\kappa = \|\mathbf{r}'(t) \times \mathbf{r}''(t)\| / \|\mathbf{r}'(t)\|^3 = 4/17$

11. $\mathbf{r}'(t) = \sinh t\mathbf{i} + \cosh t\mathbf{j} + \mathbf{k}$, $\mathbf{r}''(t) = \cosh t\mathbf{i} + \sinh t\mathbf{j}$, $\kappa = \|\mathbf{r}'(t) \times \mathbf{r}''(t)\| / \|\mathbf{r}'(t)\|^3 = \dfrac{1}{2\cosh^2 t}$

13. $\mathbf{r}'(t) = -3\sin t\mathbf{i} + 4\cos t\mathbf{j} + \mathbf{k}$, $\mathbf{r}''(t) = -3\cos t\mathbf{i} - 4\sin t\mathbf{j}$,
    $\mathbf{r}'(\pi/2) = -3\mathbf{i} + \mathbf{k}$, $\mathbf{r}''(\pi/2) = -4\mathbf{j}$; $\kappa = \|4\mathbf{i} + 12\mathbf{k}\| / \| -3\mathbf{i} + \mathbf{k}\|^3 = 2/5, \rho = 5/2$

15. $\mathbf{r}'(t) = e^t(\cos t - \sin t)\mathbf{i} + e^t(\cos t + \sin t)\mathbf{j} + e^t\mathbf{k}$,
    $\mathbf{r}''(t) = -2e^t \sin t\mathbf{i} + 2e^t \cos t\mathbf{j} + e^t\mathbf{k}$, $\mathbf{r}'(0) = \mathbf{i} + \mathbf{j} + \mathbf{k}$,
    $\mathbf{r}''(0) = 2\mathbf{j} + \mathbf{k}$; $\kappa = \| -\mathbf{i} - \mathbf{j} + 2\mathbf{k}\| / \|\mathbf{i} + \mathbf{j} + \mathbf{k}\|^3 = \sqrt{2}/3, \rho = 3\sqrt{2}/2$

17. $\mathbf{r}'(s) = \dfrac{1}{2}\cos\left(1 + \dfrac{s}{2}\right)\mathbf{i} - \dfrac{1}{2}\sin\left(1 + \dfrac{s}{2}\right)\mathbf{j} + \dfrac{\sqrt{3}}{2}\mathbf{k}$, $\|\mathbf{r}'(s)\| = 1$, so
    $\dfrac{d\mathbf{T}}{ds} = -\dfrac{1}{4}\sin\left(1 + \dfrac{s}{2}\right)\mathbf{i} - \dfrac{1}{4}\cos\left(1 + \dfrac{s}{2}\right)\mathbf{j}$, $\kappa = \left\|\dfrac{d\mathbf{T}}{ds}\right\| = \dfrac{1}{4}$

19. **(a)** $\mathbf{r}' = x'\mathbf{i} + y'\mathbf{j}, \mathbf{r}'' = x''\mathbf{i} + y''\mathbf{j}$, $\|\mathbf{r}' \times \mathbf{r}''\| = |x'y'' - x''y'|$, $\kappa = \dfrac{|x'y'' - y'x''|}{(x'^2 + y'^2)^{3/2}}$

    (b) Set $x = t$, $y = f(x) = f(t)$, $x' = 1$, $x'' = 0$, $y' = \dfrac{dy}{dx}$, $y'' = \dfrac{d^2y}{dx^2}$, $\kappa = \dfrac{|d^2y/dx^2|}{(1 + (dy/dx)^2)^{3/2}}$

21. $\kappa(x) = \dfrac{|\sin x|}{(1 + \cos^2 x)^{3/2}}$, $\kappa(\pi/2) = 1$ 23. $\kappa(x) = \dfrac{2|x|^3}{(x^4 + 1)^{3/2}}$, $\kappa(1) = 1/\sqrt{2}$

25. $\kappa(x) = \dfrac{2\sec^2 x|\tan x|}{(1 + \sec^4 x)^{3/2}}$, $\kappa(\pi/4) = 4/(5\sqrt{5})$

27. $x'(t) = 2t$, $y'(t) = 3t^2$, $x''(t) = 2$, $y''(t) = 6t$,
    $x'(1/2) = 1$, $y'(1/2) = 3/4$, $x''(1/2) = 2$, $y''(1/2) = 3$; $\kappa = 96/125$

29. $x'(t) = 3e^{3t}$, $y'(t) = -e^{-t}$, $x''(t) = 9e^{3t}$, $y''(t) = e^{-t}$,
    $x'(0) = 3$, $y'(0) = -1$, $x''(0) = 9$, $y''(0) = 1$; $\kappa = 6/(5\sqrt{10})$

31. $x'(t) = 1, y'(t) = -1/t^2, x''(t) = 0, y''(t) = 2/t^3$
    $x'(1) = 1, y'(1) = -1, x''(1) = 0, y''(1) = 2; \kappa = 1/\sqrt{2}$

33. **(a)** $\kappa(x) = \dfrac{|\cos x|}{(1 + \sin^2 x)^{3/2}}$,

$\rho(x) = \dfrac{(1 + \sin^2 x)^{3/2}}{|\cos x|}$

$\rho(0) = \rho(\pi) = 1.$

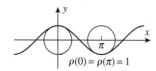

(b) $\kappa(t) = \dfrac{2}{(4\sin^2 t + \cos^2 t)^{3/2}}$,

$\rho(t) = \dfrac{1}{2}(4\sin^2 t + \cos^2 t)^{3/2}$,

$\rho(0) = 1/2, \ \rho(\pi/2) = 4$

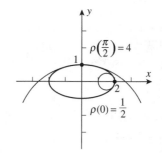

35. **(a)**

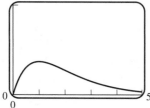

(b)

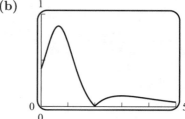

37. **(a)** $\kappa = \dfrac{|12x^2 - 4|}{\left(1 + (4x^3 - 4x)^2\right)^{3/2}}$

(b)

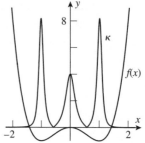

(c) $f'(x) = 4x^3 - 4x = 0$ at $x = 0, \pm 1$, $f''(x) = 12x^2 - 4$, so extrema at $x = 0, \pm 1$, and $\rho = 1/4$ for $x = 0$ and $\rho = 1/8$ when $x = \pm 1$.

39. $\mathbf{r}'(\theta) = \left(-r\sin\theta + \cos\theta \dfrac{dr}{d\theta}\right)\mathbf{i} + \left(r\cos\theta + \sin\theta \dfrac{dr}{d\theta}\right)\mathbf{j};$

$\mathbf{r}''(\theta) = \left(-r\cos\theta - 2\sin\theta \dfrac{dr}{d\theta} + \cos\theta \dfrac{d^2r}{d\theta^2}\right)\mathbf{i} + \left(-r\sin\theta + 2\cos\theta \dfrac{dr}{d\theta} + \sin\theta \dfrac{d^2r}{d\theta^2}\right)\mathbf{j};$

$\kappa = \dfrac{\left|r^2 + 2\left(\dfrac{dr}{d\theta}\right)^2 - r\dfrac{d^2r}{d\theta^2}\right|}{\left[r^2 + \left(\dfrac{dr}{d\theta}\right)^2\right]^{3/2}}.$

41. $\kappa(\theta) = \dfrac{3}{2\sqrt{2}(1 + \cos\theta)^{1/2}}$, $\kappa(\pi/2) = \dfrac{3}{2\sqrt{2}}$

43. $\kappa(\theta) = \dfrac{10 + 8\cos^2 3\theta}{(1 + 8\cos^2\theta)^{3/2}}, \ \kappa(0) = \dfrac{2}{3}$

45. The radius of curvature is zero when $\theta = \pi$, so there is a cusp there.

47. Let $y = t$, then $x = \dfrac{t^2}{4p}$ and $\kappa(t) = \dfrac{1/|2p|}{[t^2/(4p^2) + 1]^{3/2}}$;

$t = 0$ when $(x, y) = (0, 0)$ so $\kappa(0) = 1/|2p|, \ \rho = 2|p|$.

49. Let $x = 3\cos t, \ y = 2\sin t$ for $0 \le t < 2\pi, \ \kappa(t) = \dfrac{6}{(9\sin^2 t + 4\cos^2 t)^{3/2}}$ so

$\rho(t) = \dfrac{1}{6}(9\sin^2 t + 4\cos^2 t)^{3/2} = \dfrac{1}{6}(5\sin^2 t + 4)^{3/2}$ which, by inspection, is minimum when

$t = 0$ or π. The radius of curvature is minimum at $(3, 0)$ and $(-3, 0)$.

51. $\mathbf{r}'(t) = -\sin t\,\mathbf{i} + \cos t\,\mathbf{j} - \sin t\,\mathbf{k}, \ \mathbf{r}''(t) = -\cos t\,\mathbf{i} - \sin t\,\mathbf{j} - \cos t\,\mathbf{k}$,

$\|\mathbf{r}'(t) \times \mathbf{r}''(t)\| = \|-\mathbf{i} + \mathbf{k}\| = \sqrt{2}, \ \|\mathbf{r}'(t)\| = (1 + \sin^2 t)^{1/2}; \ \kappa(t) = \sqrt{2}/(1 + \sin^2 t)^{3/2}$,

$\rho(t) = (1 + \sin^2 t)^{3/2}/\sqrt{2}$. The minimum value of ρ is $1/\sqrt{2}$; the maximum value is 2.

53. From Exercise 39: $dr/d\theta = ae^{a\theta} = ar, \ d^2r/d\theta^2 = a^2 e^{a\theta} = a^2 r; \ \kappa = 1/[\sqrt{1 + a^2}\,r]$.

55. **(a)** $d^2y/dx^2 = 2, \ \kappa(\phi) = |2\cos^3\phi|$

     (b) $dy/dx = \tan\phi = 1, \phi = \pi/4, \ \kappa(\pi/4) = |2\cos^3(\pi/4)| = 1/\sqrt{2}, \ \rho = \sqrt{2}$

     (c)

57. $\kappa = 0$ along $y = 0$; along $y = x^2, \ \kappa(x) = 2/(1 + 4x^2)^{3/2}, \ \kappa(0) = 2$. Along $y = x^3$,

$\kappa(x) = 6|x|/(1 + 9x^4)^{3/2}, \ \kappa(0) = 0$.

59. $\kappa = 1/r$ along the circle; along $y = ax^2, \ \kappa(x) = 2a/(1 + 4a^2 x^2)^{3/2}, \ \kappa(0) = 2a$ so $2a = 1/r$, $a = 1/(2r)$.

61. $\kappa(x) = \dfrac{|y''|}{(1 + y'^2)^{3/2}}$ so the transition will be smooth if the values of y are equal, the values of y' are equal, and the values of y'' are equal at $x = 0$. Let $f(x)$ denote the function $y(x)$ for $x \le 0$, and set $g(x) = ax^2 + bx + c$. Then from the left $y(0) = f(0), y'(0) = f'(0), y''(0) = f''(0)$, and from the right $y(0) = g(0) = c, y'(0) = g'(0) = b$ and $y''(0) = g''(0) = 2a$. If we set $c = f(0), b = f'(0), a = f''(0)/2$ then the transition is smooth.

63. The result follows from the definitions $\mathbf{N} = \dfrac{\mathbf{T}'(s)}{\|\mathbf{T}'(s)\|}$ and $\kappa = \|\mathbf{T}'(s)\|$.

65. $\dfrac{d\mathbf{N}}{ds} = \mathbf{B} \times \dfrac{d\mathbf{T}}{ds} + \dfrac{d\mathbf{B}}{ds} \times \mathbf{T} = \mathbf{B} \times (\kappa\mathbf{N}) + (-\tau\mathbf{N}) \times \mathbf{T} = \kappa\mathbf{B} \times \mathbf{N} - \tau\mathbf{N} \times \mathbf{T}$, but $\mathbf{B} \times \mathbf{N} = -\mathbf{T}$ and

$\mathbf{N} \times \mathbf{T} = -\mathbf{B}$ so $\dfrac{d\mathbf{N}}{ds} = -\kappa\mathbf{T} + \tau\mathbf{B}$

67. $\mathbf{r} = a\cos(s/w)\mathbf{i} + a\sin(s/w)\mathbf{j} + (cs/w)\mathbf{k}$, $\quad \mathbf{r}' = -(a/w)\sin(s/w)\mathbf{i} + (a/w)\cos(s/w)\mathbf{j} + (c/w)\mathbf{k}$,

$\mathbf{r}'' = -(a/w^2)\cos(s/w)\mathbf{i} - (a/w^2)\sin(s/w)\mathbf{j}$, $\quad \mathbf{r}''' = (a/w^3)\sin(s/w)\mathbf{i} - (a/w^3)\cos(s/w)\mathbf{j}$,

$\mathbf{r}' \times \mathbf{r}'' = (ac/w^3)\sin(s/w)\mathbf{i} - (ac/w^3)\cos(s/w)\mathbf{j} + (a^2/w^3)\mathbf{k}$, $\quad (\mathbf{r}' \times \mathbf{r}'') \cdot \mathbf{r}''' = a^2c/w^6$,

$\|\mathbf{r}''(s)\| = a/w^2$, so $\tau = c/w^2$ and $\mathbf{B} = (c/w)\sin(s/w)\mathbf{i} - (c/w)\cos(s/w)\mathbf{j} + (a/w)\mathbf{k}$

69. $\mathbf{r}' = 2\mathbf{i} + 2t\mathbf{j} + t^2\mathbf{k}$, $\mathbf{r}'' = 2\mathbf{j} + 2t\mathbf{k}$, $\mathbf{r}''' = 2\mathbf{k}$, $\mathbf{r}' \times \mathbf{r}'' = 2t^2\mathbf{i} - 4t\mathbf{j} + 4\mathbf{k}$, $\|\mathbf{r}' \times \mathbf{r}''\| = 2(t^2 + 2)$,

$\tau = 8/[2(t^2 + 2)]^2 = 2/(t^2 + 2)^2$

71. $\mathbf{r}' = e^t\mathbf{i} - e^{-t}\mathbf{j} + \sqrt{2}\mathbf{k}$, $\mathbf{r}'' = e^t\mathbf{i} + e^{-t}\mathbf{j}$, $\mathbf{r}''' = e^t\mathbf{i} - e^{-t}\mathbf{j}$, $\mathbf{r}' \times \mathbf{r}'' = -\sqrt{2}e^{-t}\mathbf{i} + \sqrt{2}e^t\mathbf{j} + 2\mathbf{k}$,

$\|\mathbf{r}' \times \mathbf{r}''\| = \sqrt{2}(e^t + e^{-t})$, $\tau = (-2\sqrt{2})/[2(e^t + e^{-t})^2] = -\sqrt{2}/(e^t + e^{-t})^2$

EXERCISE SET 13.6

1. $\mathbf{v}(t) = -3\sin t\mathbf{i} + 3\cos t\mathbf{j}$

$\mathbf{a}(t) = -3\cos t\mathbf{i} - 3\sin t\mathbf{j}$

$\|\mathbf{v}(t)\| = \sqrt{9\sin^2 t + 9\cos^2 t} = 3$

$\mathbf{r}(\pi/3) = (3/2)\mathbf{i} + (3\sqrt{3}/2)\mathbf{j}$

$\mathbf{v}(\pi/3) = -(3\sqrt{3}/2)\mathbf{i} + (3/2)\mathbf{j}$

$\mathbf{a}(\pi/3) = -(3/2)\mathbf{i} - (3\sqrt{3}/2)\mathbf{j}$

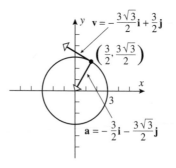

3. $\mathbf{v}(t) = e^t\mathbf{i} - e^{-t}\mathbf{j}$

$\mathbf{a}(t) = e^t\mathbf{i} + e^{-t}\mathbf{j}$

$\|\mathbf{v}(t)\| = \sqrt{e^{2t} + e^{-2t}}$

$\mathbf{r}(0) = \mathbf{i} + \mathbf{j}$

$\mathbf{v}(0) = \mathbf{i} - \mathbf{j}$

$\mathbf{a}(0) = \mathbf{i} + \mathbf{j}$

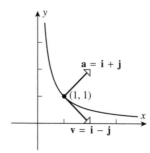

5. $\mathbf{v} = \mathbf{i} + t\mathbf{j} + t^2\mathbf{k}$, $\mathbf{a} = \mathbf{j} + 2t\mathbf{k}$; at $t = 1$, $\mathbf{v} = \mathbf{i} + \mathbf{j} + \mathbf{k}$, $\|\mathbf{v}\| = \sqrt{3}$, $\mathbf{a} = \mathbf{j} + 2\mathbf{k}$

7. $\mathbf{v} = -2\sin t\mathbf{i} + 2\cos t\mathbf{j} + \mathbf{k}$, $\mathbf{a} = -2\cos t\mathbf{i} - 2\sin t\mathbf{j}$;

at $t = \pi/4$, $\mathbf{v} = -\sqrt{2}\mathbf{i} + \sqrt{2}\mathbf{j} + \mathbf{k}$, $\|\mathbf{v}\| = \sqrt{5}$, $\mathbf{a} = -\sqrt{2}\mathbf{i} - \sqrt{2}\mathbf{j}$

9. **(a)** $\mathbf{v} = -a\omega\sin\omega t\mathbf{i} + b\omega\cos\omega t\mathbf{j}$, $\mathbf{a} = -a\omega^2\cos\omega t\mathbf{i} - b\omega^2\sin\omega t\mathbf{j} = -\omega^2\mathbf{r}$

(b) From Part (a), $\|\mathbf{a}\| = \omega^2\|\mathbf{r}\|$

11. If $\mathbf{a} = \mathbf{0}$ then $x''(t) = y''(t) = z''(t) = 0$, so $x(t) = x_1t + x_0, y(t) = y_1t + y_0, z(t) = z_1t + z_0$, the motion is along a straight line and has constant speed.

13. $\mathbf{v} = (6/\sqrt{t})\mathbf{i} + (3/2)t^{1/2}\mathbf{j}$, $\|\mathbf{v}\| = \sqrt{36/t + 9t/4}$, $d\|\mathbf{v}\|/dt = (-36/t^2 + 9/4)/(2\sqrt{36/t + 9t/4}) = 0$ if $t = 4$ which yields a minimum by the first derivative test. The minimum speed is $3\sqrt{2}$ when $\mathbf{r} = 24\mathbf{i} + 8\mathbf{j}$.

15. (a)

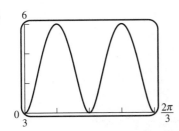

(b) $\mathbf{v} = 3\cos 3t\mathbf{i} + 6\sin 3t\mathbf{j}$, $\|\mathbf{v}\| = \sqrt{9\cos^2 3t + 36\sin^2 3t} = 3\sqrt{1 + 3\sin^2 3t}$; by inspection, maximum speed is 6 and minimum speed is 3

(d) $\dfrac{d}{dt}\|\mathbf{v}\| = \dfrac{27\sin 6t}{2\sqrt{1 + 3\sin^2 3t}} = 0$ when $t = 0, \pi/6, \pi/3, \pi/2, 2\pi/3$; the maximum speed is 6 which occurs first when $\sin 3t = 1, t = \pi/6$.

17. $\mathbf{v}(t) = -\sin t\mathbf{i} + \cos t\mathbf{j} + \mathbf{C}_1$, $\mathbf{v}(0) = \mathbf{j} + \mathbf{C}_1 = \mathbf{i}$, $\mathbf{C}_1 = \mathbf{i} - \mathbf{j}$, $\mathbf{v}(t) = (1 - \sin t)\mathbf{i} + (\cos t - 1)\mathbf{j}$;
$\mathbf{r}(t) = (t + \cos t)\mathbf{i} + (\sin t - t)\mathbf{j} + \mathbf{C}_2$, $\mathbf{r}(0) = \mathbf{i} + \mathbf{C}_2 = \mathbf{j}$,
$\mathbf{C}_2 = -\mathbf{i} + \mathbf{j}$ so $\mathbf{r}(t) = (t + \cos t - 1)\mathbf{i} + (\sin t - t + 1)\mathbf{j}$

19. $\mathbf{v}(t) = -\cos t\mathbf{i} + \sin t\mathbf{j} + e^t\mathbf{k} + \mathbf{C}_1$, $\mathbf{v}(0) = -\mathbf{i} + \mathbf{k} + \mathbf{C}_1 = \mathbf{k}$ so
$\mathbf{C}_1 = \mathbf{i}$, $\mathbf{v}(t) = (1 - \cos t)\mathbf{i} + \sin t\mathbf{j} + e^t\mathbf{k}$; $\mathbf{r}(t) = (t - \sin t)\mathbf{i} - \cos t\mathbf{j} + e^t\mathbf{k} + \mathbf{C}_2$,
$\mathbf{r}(0) = -\mathbf{j} + \mathbf{k} + \mathbf{C}_2 = -\mathbf{i} + \mathbf{k}$ so $\mathbf{C}_2 = -\mathbf{i} + \mathbf{j}$, $\mathbf{r}(t) = (t - \sin t - 1)\mathbf{i} + (1 - \cos t)\mathbf{j} + e^t\mathbf{k}$.

21. $\mathbf{v} = 3t^2\mathbf{i} + 2t\mathbf{j}$, $\mathbf{a} = 6t\mathbf{i} + 2\mathbf{j}$; $\mathbf{v} = 3\mathbf{i} + 2\mathbf{j}$ and $\mathbf{a} = 6\mathbf{i} + 2\mathbf{j}$ when $t = 1$ so
$\cos\theta = (\mathbf{v} \cdot \mathbf{a})/(\|\mathbf{v}\|\,\|\mathbf{a}\|) = 11/\sqrt{130}$, $\theta \approx 15°$.

23. (a) displacement $= \mathbf{r}_1 - \mathbf{r}_0 = 0.7\mathbf{i} + 2.7\mathbf{j} - 3.4\mathbf{k}$

    (b) $\Delta\mathbf{r} = \mathbf{r}_1 - \mathbf{r}_0$, so $\mathbf{r}_0 = \mathbf{r}_1 - \Delta\mathbf{r} = -0.7\mathbf{i} - 2.9\mathbf{j} + 4.8\mathbf{k}$.

25. $\Delta\mathbf{r} = \mathbf{r}(3) - \mathbf{r}(1) = 8\mathbf{i} + (26/3)\mathbf{j}$; $\mathbf{v} = 2t\mathbf{i} + t^2\mathbf{j}$, $s = \displaystyle\int_1^3 t\sqrt{4 + t^2}\,dt = (13\sqrt{13} - 5\sqrt{5})/3$.

27. $\Delta\mathbf{r} = \mathbf{r}(\ln 3) - \mathbf{r}(0) = 2\mathbf{i} - (2/3)\mathbf{j} + \sqrt{2}(\ln 3)\mathbf{k}$; $\mathbf{v} = e^t\mathbf{i} - e^{-t}\mathbf{j} + \sqrt{2}\mathbf{k}$, $s = \displaystyle\int_0^{\ln 3} (e^t + e^{-t})\,dt = 8/3$.

29. In both cases, the equation of the path in rectangular coordinates is $x^2 + y^2 = 4$, the particles move counterclockwise around this circle; $\mathbf{v}_1 = -6\sin 3t\mathbf{i} + 6\cos 3t\mathbf{j}$ and $\mathbf{v}_2 = -4t\sin(t^2)\mathbf{i} + 4t\cos(t^2)\mathbf{j}$ so $\|\mathbf{v}_1\| = 6$ and $\|\mathbf{v}_2\| = 4t$.

31. (a) $\mathbf{v} = -e^{-t}\mathbf{i} + e^t\mathbf{j}$, $\mathbf{a} = e^{-t}\mathbf{i} + e^t\mathbf{j}$; when $t = 0$, $\mathbf{v} = -\mathbf{i} + \mathbf{j}$, $\mathbf{a} = \mathbf{i} + \mathbf{j}$, $\|\mathbf{v}\| = \sqrt{2}$, $\mathbf{v} \cdot \mathbf{a} = 0$, $\mathbf{v} \times \mathbf{a} = -2\mathbf{k}$ so $a_T = 0$, $a_N = \sqrt{2}$.

    (b) $a_T\mathbf{T} = \mathbf{0}$, $a_N\mathbf{N} = \mathbf{a} - a_T\mathbf{T} = \mathbf{i} + \mathbf{j}$ **(c)** $\kappa = 1/\sqrt{2}$

33. (a) $\mathbf{v} = (3t^2 - 2)\mathbf{i} + 2t\mathbf{j}$, $\mathbf{a} = 6t\mathbf{i} + 2\mathbf{j}$; when $t = 1$, $\mathbf{v} = \mathbf{i} + 2\mathbf{j}$, $\mathbf{a} = 6\mathbf{i} + 2\mathbf{j}$, $\|\mathbf{v}\| = \sqrt{5}$, $\mathbf{v} \cdot \mathbf{a} = 10$, $\mathbf{v} \times \mathbf{a} = -10\mathbf{k}$ so $a_T = 2\sqrt{5}$, $a_N = 2\sqrt{5}$

    (b) $a_T\mathbf{T} = \dfrac{2\sqrt{5}}{\sqrt{5}}(\mathbf{i} + 2\mathbf{j}) = 2\mathbf{i} + 4\mathbf{j}$, $a_N\mathbf{N} = \mathbf{a} - a_T\mathbf{T} = 4\mathbf{i} - 2\mathbf{j}$

    (c) $\kappa = 2/\sqrt{5}$

35. (a) $\mathbf{v} = (-1/t^2)\mathbf{i} + 2t\mathbf{j} + 3t^2\mathbf{k}$, $\mathbf{a} = (2/t^3)\,\mathbf{i} + 2\mathbf{j} + 6t\mathbf{k}$; when $t = 1$, $\mathbf{v} = -\mathbf{i} + 2\mathbf{j} + 3\mathbf{k}$, $\mathbf{a} = 2\mathbf{i} + 2\mathbf{j} + 6\mathbf{k}$, $\|\mathbf{v}\| = \sqrt{14}$, $\mathbf{v} \cdot \mathbf{a} = 20$, $\mathbf{v} \times \mathbf{a} = 6\mathbf{i} + 12\mathbf{j} - 6\mathbf{k}$ so $a_T = 20/\sqrt{14}$, $a_N = 6\sqrt{3}/\sqrt{7}$

   (b) $a_T \mathbf{T} = -\dfrac{10}{7}\mathbf{i} + \dfrac{20}{7}\mathbf{j} + \dfrac{30}{7}\mathbf{k}$, $a_N \mathbf{N} = \mathbf{a} - a_T \mathbf{T} = \dfrac{24}{7}\mathbf{i} - \dfrac{6}{7}\mathbf{j} + \dfrac{12}{7}\mathbf{k}$

   (c) $\kappa = \dfrac{6\sqrt{6}}{14^{3/2}} = \left(\dfrac{3}{7}\right)^{3/2}$

37. (a) $\mathbf{v} = 3\cos t\,\mathbf{i} - 2\sin t\,\mathbf{j} - 2\cos 2t\,\mathbf{k}$, $\mathbf{a} = -3\sin t\,\mathbf{i} - 2\cos t\,\mathbf{j} + 4\sin 2t\,\mathbf{k}$; when $t = \pi/2$, $\mathbf{v} = -2\mathbf{j} + 2\mathbf{k}$, $\mathbf{a} = -3\mathbf{i}$, $\|\mathbf{v}\| = 2\sqrt{2}$, $\mathbf{v} \cdot \mathbf{a} = 0$, $\mathbf{v} \times \mathbf{a} = -6\mathbf{j} - 6\mathbf{k}$ so $a_T = 0$, $a_N = 3$

   (b) $a_T \mathbf{T} = \mathbf{0}$, $a_N \mathbf{N} = \mathbf{a} = -3\mathbf{i}$

   (c) $\kappa = \dfrac{3}{8}$

39. $\|\mathbf{v}\| = 4$, $\mathbf{v} \cdot \mathbf{a} = -12$, $\mathbf{v} \times \mathbf{a} = 8\mathbf{k}$ so $a_T = -3$, $a_N = 2$, $\mathbf{T} = -\mathbf{j}$, $\mathbf{N} = (\mathbf{a} - a_T \mathbf{T})/a_N = \mathbf{i}$

41. $\|\mathbf{v}\| = 3$, $\mathbf{v} \cdot \mathbf{a} = 4$, $\mathbf{v} \times \mathbf{a} = 4\mathbf{i} - 3\mathbf{j} - 2\mathbf{k}$ so $a_T = 4/3$, $a_N = \sqrt{29}/3$, $\mathbf{T} = (1/3)(2\mathbf{i} + 2\mathbf{j} + \mathbf{k})$, $\mathbf{N} = (\mathbf{a} - a_T \mathbf{T})/a_N = (\mathbf{i} - 8\mathbf{j} + 14\mathbf{k})/(3\sqrt{29})$

43. $a_T = \dfrac{d^2 s}{dt^2} = \dfrac{d}{dt}\sqrt{3t^2 + 4} = 3t/\sqrt{3t^2 + 4}$ so when $t = 2$, $a_T = 3/2$.

45. $a_T = \dfrac{d^2 s}{dt^2} = \dfrac{d}{dt}\sqrt{(4t-1)^2 + \cos^2 \pi t} = [4(4t-1) - \pi \cos \pi t \sin \pi t]/\sqrt{(4t-1)^2 + \cos^2 \pi t}$ so when $t = 1/4$, $a_T = -\pi/\sqrt{2}$.

47. $a_N = \kappa (ds/dt)^2 = (1/\rho)(ds/dt)^2 = (1/1)(2.9 \times 10^5)^2 = 8.41 \times 10^{10}$ km/s^2

49. $a_N = \kappa (ds/dt)^2 = [2/(1 + 4x^2)^{3/2}](3)^2 = 18/(1 + 4x^2)^{3/2}$

51. $\mathbf{a} = a_T \mathbf{T} + a_N \mathbf{N}$; by the Pythagorean Theorem $a_N = \sqrt{\|\mathbf{a}\|^2 - a_T^2} = \sqrt{9 - 9} = 0$

53. Let $c = ds/dt$, $a_N = \kappa \left(\dfrac{ds}{dt}\right)^2$, $a_N = \dfrac{1}{1000}c^2$, so $c^2 = 1000 a_N$, $c \le 10\sqrt{10}\sqrt{1.5} \approx 38.73$ m/s.

55. (a) $v_0 = 320$, $\alpha = 60°$, $s_0 = 0$ so $x = 160t$, $y = 160\sqrt{3}t - 16t^2$.

   (b) $dy/dt = 160\sqrt{3} - 32t$, $dy/dt = 0$ when $t = 5\sqrt{3}$ so $y_{\max} = 160\sqrt{3}(5\sqrt{3}) - 16(5\sqrt{3})^2 = 1200$ ft.

   (c) $y = 16t(10\sqrt{3} - t)$, $y = 0$ when $t = 0$ or $10\sqrt{3}$ so $x_{\max} = 160(10\sqrt{3}) = 1600\sqrt{3}$ ft.

   (d) $\mathbf{v}(t) = 160\mathbf{i} + (160\sqrt{3} - 32t)\mathbf{j}$, $\mathbf{v}(10\sqrt{3}) = 160(\mathbf{i} - \sqrt{3}\mathbf{j})$, $\|\mathbf{v}(10\sqrt{3})\| = 320$ ft/s.

57. $v_0 = 80$, $\alpha = -60°$, $s_0 = 168$ so $x = 40t$, $y = 168 - 40\sqrt{3}\,t - 16t^2$; $y = 0$ when $t = -7\sqrt{3}/2$ (invalid) or $t = \sqrt{3}$ so $x(\sqrt{3}) = 40\sqrt{3}$ ft.

59. $\alpha = 30°$, $s_0 = 0$ so $x = \sqrt{3}v_0 t/2$, $y = v_0 t/2 - 16t^2$; $dy/dt = v_0/2 - 32t$, $dy/dt = 0$ when $t = v_0/64$ so $y_{\max} = v_0^2/256 = 2500$, $v_0 = 800$ ft/s.

61. $v_0 = 800, s_0 = 0$ so $x = (800 \cos \alpha)t$, $y = (800 \sin \alpha)t - 16t^2 = 16t(50 \sin \alpha - t)$; $y = 0$ when $t = 0$ or $50 \sin \alpha$ so $x_{\max} = 40,000 \sin \alpha \cos \alpha = 20,000 \sin 2\alpha = 10,000$, $2\alpha = 30°$ or $150°$, $\alpha = 15°$ or $75°$.

63. **(a)** Let $\mathbf{r}(t) = x(t)\mathbf{i} + y(t)\mathbf{j}$ with $\mathbf{j}$ pointing up. Then $\mathbf{a} = -32\mathbf{j} = x''(t)\mathbf{i} + y''(t)\mathbf{j}$, so $x(t) = At + B, y(t) = -16t^2 + Ct + D$. Next, $x(0) = 0, y(0) = 4$ so $x(t) = At, y(t) = -16t^2 + Ct + 4$; $y'(0)/x'(0) = \tan 60° = \sqrt{3}$, so $C = \sqrt{3}A$; and $40 = v_0 = \sqrt{x'(0)^2 + y'(0)^2} = \sqrt{A^2 + 3A^2}, A = 20$, thus $\mathbf{r}(t) = 20t\,\mathbf{i} + (-16t^2 + 20\sqrt{3}t + 4)\,\mathbf{j}$. When $x = 15$, $t = \dfrac{3}{4}$, and $y = 4 + 20\sqrt{3}\dfrac{3}{4} - 16\left(\dfrac{3}{4}\right)^2 \approx 20.98$ ft, so the water clears the corner point A with 0.98 ft to spare.

(b) $y = 20$ when $-16t^2 + 20\sqrt{3}t - 16 = 0, t = 0.668$ (reject) or $1.497, x(1.497) \approx 29.942$ ft, so the water hits the roof.

(c) about $29.942 - 15 = 14.942$ ft

65. **(a)** $x = (35\sqrt{2}/2)t$, $y = (35\sqrt{2}/2)t - 4.9t^2$, from Exercise 19a in Section 13.5
$$\kappa = \frac{|x'y'' - x''y'|}{[(x')^2 + (y')^2]^{3/2}}, \ \kappa(0) = \frac{9.8}{35^2\sqrt{2}} = 0.004\sqrt{2} \approx 0.00565685; \rho = 1/\kappa \approx 176.78 \text{ m}$$

(b) $y'(t) = 0$ when $t = \dfrac{25}{14}\sqrt{2}, y = \dfrac{125}{4}$ m

67. $s_0 = 0$ so $x = (v_0 \cos \alpha)t$, $y = (v_0 \sin \alpha)t - gt^2/2$

(a) $dy/dt = v_0 \sin \alpha - gt$ so $dy/dt = 0$ when $t = (v_0 \sin \alpha)/g$, $y_{\max} = (v_0 \sin \alpha)^2/(2g)$

(b) $y = 0$ when $t = 0$ or $(2v_0 \sin \alpha)/g$, so $x = R = (2v_0^2 \sin \alpha \cos \alpha)/g = (v_0^2 \sin 2\alpha)/g$ when $t = (2v_0 \sin \alpha)/g$; R is maximum when $2\alpha = 90°$, $\alpha = 45°$, and the maximum value of R is v_0^2/g.

69. $v_0 = 80$, $\alpha = 30°$, $s_0 = 5$ so $x = 40\sqrt{3}t$, $y = 5 + 40t - 16t^2$

(a) $y = 0$ when $t = (-40 \pm \sqrt{(40)^2 - 4(-16)(5)})/(-32) = (5 \pm \sqrt{30})/4$, reject $(5 - \sqrt{30})/4$ to get $t = (5 + \sqrt{30})/4 \approx 2.62$ s.

(b) $x \approx 40\sqrt{3}(2.62) \approx 181.5$ ft.

71. **(a)** $v_0(\cos \alpha)(2.9) = 259 \cos 23°$ so $v_0 \cos \alpha \approx 82.21061$, $v_0(\sin \alpha)(2.9) - 16(2.9)^2 = -259 \sin 23°$ so $v_0 \sin \alpha \approx 11.50367$; divide $v_0 \sin \alpha$ by $v_0 \cos \alpha$ to get $\tan \alpha \approx 0.139929$, thus $\alpha \approx 8°$ and $v_0 \approx 82.21061/\cos 8° \approx 83$ ft/s.

(b) From Part (a), $x \approx 82.21061t$ and $y \approx 11.50367t - 16t^2$ for $0 \le t \le 2.9$; the distance traveled is $\displaystyle\int_0^{2.9} \sqrt{(dx/dt)^2 + (dy/dt)^2}\,dt \approx 268.76$ ft.

73. **(a)** $\|\mathbf{e}_r(t)\|^2 = \cos^2 \theta + \sin^2 \theta = 1$, so $\mathbf{e}_r(t)$ is a unit vector; $\mathbf{r}(t) = r(t)\mathbf{e}(t)$, so they have the same direction if $r(t) > 0$, opposite if $r(t) < 0$. $\mathbf{e}_\theta(t)$ is perpendicular to $\mathbf{e}_r(t)$ since $\mathbf{e}_r(t) \cdot \mathbf{e}_\theta(t) = 0$, and it will result from a counterclockwise rotation of $\mathbf{e}_r(t)$ provided $\mathbf{e}(t) \times \mathbf{e}_\theta(t) = \mathbf{k}$, which is true.

(b) $\dfrac{d}{dt}\mathbf{e}_r(t) = \dfrac{d\theta}{dt}(-\sin\theta\mathbf{i} + \cos\theta\mathbf{j}) = \dfrac{d\theta}{dt}\mathbf{e}_\theta(t)$ and $\dfrac{d}{dt}\mathbf{e}_\theta(t) = -\dfrac{d\theta}{dt}(\cos\theta\mathbf{i} + \sin\theta\mathbf{j}) = -\dfrac{d\theta}{dt}\mathbf{e}_r(t)$, so

$$\mathbf{v}(t) = \frac{d}{dt}\mathbf{r}(t) = \frac{d}{dt}(r(t)\mathbf{e}_r(t)) = r'(t)\mathbf{e}_r(t) + r(t)\frac{d\theta}{dt}\mathbf{e}_\theta(t)$$

(c) From Part (b),

$$\mathbf{a} = \frac{d}{dt}\mathbf{v}(t)$$

$$= r''(t)\mathbf{e}_r(t) + r'(t)\frac{d\theta}{dt}\mathbf{e}_\theta(t) + r'(t)\frac{d\theta}{dt}\mathbf{e}_\theta(t) + r(t)\frac{d^2\theta}{dt^2}\mathbf{e}_\theta(t) - r(t)\left(\frac{d\theta}{dt}\right)^2\mathbf{e}_r(t)$$

$$= \left[\frac{d^2 r}{dt^2} - r\left(\frac{d\theta}{dt}\right)^2\right]\mathbf{e}_r(t) + \left[r\frac{d^2\theta}{dt^2} + 2\frac{dr}{dt}\frac{d\theta}{dt}\right]\mathbf{e}_\theta(t)$$

EXERCISE SET 13.7

1. **(a)** From (15) and (6), at $t = 0$,
$$\mathbf{C} = \mathbf{v}_0 \times \mathbf{b}_0 - GM\mathbf{u} = v_0\mathbf{j} \times r_0 v_0\mathbf{k} - GM\mathbf{u} = r_0 v_0^2\mathbf{i} - GM\mathbf{i} = (r_0 v_0^2 - GM)\mathbf{i}$$

   (b) From (22), $r_0 v_0^2 - GM = GMe$, so from (7) and (17), $\mathbf{v} \times \mathbf{b} = GM(\cos\theta\mathbf{i} + \sin\theta\mathbf{j}) + GMe\mathbf{i}$, and the result follows.

   (c) From (10) it follows that $\mathbf{b}$ is perpendicular to $\mathbf{v}$, and the result follows.

   (d) From Part (c) and (10), $\|\mathbf{v} \times \mathbf{b}\| = \|\mathbf{v}\|\|\mathbf{b}\| = vr_0 v_0$. From Part (b),
   $\|\mathbf{v} \times \mathbf{b}\| = GM\sqrt{(e + \cos\theta)^2 + \sin^2\theta} = GM\sqrt{e^2 + 2e\cos\theta + 1}$. By (10) and
   Part (c), $\|\mathbf{v} \times \mathbf{b}\| = \|\mathbf{v}\|\|\mathbf{b}\| = v(r_0 v_0)$ thus $v = \dfrac{GM}{r_0 v_0}\sqrt{e^2 + 2e\cos\theta + 1}$. From (22),
   $r_0 v_0^2/(GM) = 1 + e$, $GM/(r_0 v_0) = v_0/(1 + e)$ so $v = \dfrac{v_0}{1 + e}\sqrt{e^2 + 2e\cos\theta + 1}$.

   (e) From (20) $r = \dfrac{k}{1 + e\cos\theta}$, so the minimum value of r occurs when $\theta = 0$ and the maximum
   value when $\theta = \pi$. From Part (d) $v = \dfrac{v_0}{1 + e}\sqrt{e^2 + 2e\cos\theta + 1}$ so the minimum value of v
   occurs when $\theta = \pi$ and the maximum when $\theta = 0$

3. $v_{\max}$ occurs when $\theta = 0$ so $v_{\max} = v_0$; $v_{\min}$ occurs when $\theta = \pi$ so
$v_{\min} = \dfrac{v_0}{1 + e}\sqrt{e^2 - 2e + 1} = v_{\max}\dfrac{1 - e}{1 + e}$, thus $v_{\max} = v_{\min}\dfrac{1 + e}{1 - e}$.

5. **(a)** The results follow from formulae (1) and (7) of Section 11.6.

   (b) $r_{\min}$ and $r_{\max}$ are extremes and occur at the same time as the extrema of $\|\mathbf{r}\|^2$, and hence
   at critical points of $\|\mathbf{r}\|^2$. Thus $\dfrac{d}{dt}\|\mathbf{r}\|^2 = \dfrac{d}{dt}(\mathbf{r}\cdot\mathbf{r}) = 2\mathbf{r}\cdot\mathbf{r}' = 0$, and hence $\mathbf{r}$ and $\mathbf{v} = \mathbf{r}'$ are
   orthogonal.

   (c) $v_{\min}$ and $v_{\max}$ are extremes and occur at the same time as the extrema of $\|\mathbf{v}\|^2$, and hence
   at critical points of $\|\mathbf{v}\|^2$. Thus $\dfrac{d}{dt}\|\mathbf{v}\|^2 = \dfrac{d}{dt}(\mathbf{v}\cdot\mathbf{v}) = 2\mathbf{v}\cdot\mathbf{v}' = 0$, and hence $\mathbf{v}$ and $\mathbf{a} = \mathbf{v}'$
   are orthogonal. By (5), $\mathbf{a}$ is a scalar multiple of $\mathbf{r}$ and thus $\mathbf{v}$ and $\mathbf{r}$ are orthogonal.

   (d) From equation (2) it follows that $\mathbf{r} \times \mathbf{v} = \mathbf{b}$ and thus $\|\mathbf{b}\| = \|\mathbf{r} \times \mathbf{v}\| = \|\mathbf{r}\|\,\|\mathbf{v}\|\sin\theta$. When
   either $\mathbf{r}$ or $\mathbf{v}$ has an extremum, however, the angle $\theta = 0$ and thus $\|\mathbf{b}\| = \|\mathbf{r}\|\|\mathbf{v}\|$. Finally,
   since $\mathbf{b}$ is a constant vector, the maximum of $\mathbf{r}$ occurs at the minimum of $\mathbf{v}$ and vice versa,
   and thus $\|\mathbf{b}\| = r_{\max}v_{\min} = r_{\min}v_{\max}$.

7. $r_0 = 6440 + 200 = 6640$ km so $v = \sqrt{3.99 \times 10^5/6640} \approx 7.75$ km/s.

9. From (23) with $r_0 = 6440 + 300 = 6740$ km, $v_{esc} = \sqrt{\dfrac{2(3.99) \times 10^5}{6740}} \approx 10.88$ km/s.

11. (a) At perigee, $r = r_{min} = a(1 - e) = 238{,}900\ (1 - 0.055) \approx 225{,}760$ mi; at apogee, $r = r_{max} = a(1 + e) = 238{,}900(1 + 0.055) \approx 252{,}040$ mi. Subtract the sum of the radius of the Moon and the radius of the Earth to get minimum distance $= 225{,}760 - 5080 = 220{,}680$ mi, and maximum distance $= 252{,}040 - 5080 = 246{,}960$ mi.

    (b) $T = 2\pi\sqrt{a^3/(GM)} = 2\pi\sqrt{(238{,}900)^3/(1.24 \times 10^{12})} \approx 659$ hr ≈ 27.5 days.

13. (a) $r_0 = 4000 + 180 = 4180$ mi, $v = \sqrt{\dfrac{GM}{r_0}} = \sqrt{1.24 \times 10^{12}/4180} \approx 17{,}224$ mi/h

    (b) $r_0 = 4180$ mi, $v_0 = \sqrt{\dfrac{GM}{r_0}} + 600$; $e = \dfrac{r_0 v_0^2}{GM} - 1 = 1200\sqrt{\dfrac{r_0}{GM}} + (600)^2\dfrac{r_0}{GM} \approx 0.071$; $r_{max} = 4180(1 + 0.071)/(1 - 0.071) \approx 4819$ mi; the apogee altitude is $4819 - 4000 = 819$ mi.

REVIEW EXERCISES, CHAPTER 13

3. the circle of radius 3 in the xy-plane, with center at the origin

5. a parabola in the plane $x = -2$, vertex at $(-2, 0, -1)$, opening upward

7. Let $\mathbf{r} = x\mathbf{i} + y\mathbf{j} + z\mathbf{k}$, then $x^2 + z^2 = t^2(\sin^2 \pi t + \cos^2 \pi t) = t^2 = y^2$

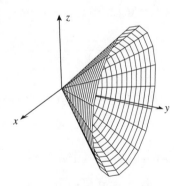

11. $\mathbf{r}'(t) = (1 - 2\sin 2t)\mathbf{i} - (2t + 1)\mathbf{j} + \cos t\mathbf{k}, \mathbf{r}'(0) = \mathbf{i} - \mathbf{j} + \mathbf{k}$ and $\mathbf{r}(0) = \mathbf{i}$, so the equation of the line is $\mathbf{r}(t) = \mathbf{i} + t(\mathbf{i} - \mathbf{j} + \mathbf{k}) = (1 + t)\mathbf{i} - t\mathbf{j} + t\mathbf{k}$.

13. $(\sin t)\mathbf{i} - (\cos t)\mathbf{j} + \mathbf{C}$

15. $\mathbf{y}(t) = \displaystyle\int \mathbf{y}'(t)dt = \tfrac{1}{3}t^3\mathbf{i} + t^2\mathbf{j} + \mathbf{C}, \mathbf{y}(0) = \mathbf{C} = \mathbf{i} + \mathbf{j}, \mathbf{y}(t) = (\tfrac{1}{3}t^3 + 1)\mathbf{i} + (t^2 + 1)\mathbf{j}$

17. $\left(\dfrac{ds}{dt}\right)^2 = \left(\sqrt{2}e^{\sqrt{2}t}\right)^2 + \left(-\sqrt{2}e^{-\sqrt{2}t}\right)^2 + 4 = 8\cosh^2(\sqrt{2}t),$

$$L = \int_0^{\sqrt{2}\ln 2} 2\sqrt{2}\cosh(\sqrt{2}t)\,dt = 2\sinh(\sqrt{2}t)\Big]_0^{\sqrt{2}\ln 2} = 2\sinh(2\ln 2) = \dfrac{15}{4}$$

19. $\mathbf{r} = \mathbf{r}_0 + t\,\overrightarrow{PQ} = (t-1)\mathbf{i} + (4-2t)\mathbf{j} + (3+2t)\mathbf{k};\; \left\|\dfrac{d\mathbf{r}}{dt}\right\| = 3, \mathbf{r}(s) = \dfrac{s-3}{3}\mathbf{i} + \dfrac{12-2s}{3}\mathbf{j} + \dfrac{9+2s}{3}\mathbf{k}$

25. $\mathbf{r}'(t) = -2\sin t\,\mathbf{i} + 3\cos t\,\mathbf{j} - \mathbf{k}, \mathbf{r}'(\pi/2) = -2\mathbf{i} - \mathbf{k}$

$\mathbf{r}''(t) = -2\cos t\,\mathbf{i} - 3\sin t\,\mathbf{j}, \mathbf{r}''(\pi/2) = -3\mathbf{j},$

$\mathbf{r}'(\pi/2) \times \mathbf{r}''(\pi/2) = -3\mathbf{i} + 6\mathbf{k}$ and hence by Theorem 13.5.2b,

$\kappa(\pi/2) = \sqrt{45}/5^{3/2} = 3/5.$

27. By Exercise 19(b) of Section 13.5, $\kappa = |d^2y/dx^2|/[1 + (dy/dx)^2]^{3/2}$, but $d^2y/dx^2 = -\cos x$ and at $x = \pi/2, d^2y/dx^2 = 0$, so $\kappa = 0$.

29. **(a)** speed   **(b)** distance traveled   **(c)** distance of the particle from the origin

31. **(a)** $\|\mathbf{r}(t)\| = 1$, so by Theorem 13.2.9, $\mathbf{r}'(t)$ is always perpendicular to the vector $\mathbf{r}(t)$. Then $\mathbf{v}(t) = R\omega(-\sin\omega t\,\mathbf{i} + \cos\omega t\,\mathbf{j}), v = \|\mathbf{v}(t)\| = R\omega$

(b) $\mathbf{a} = -R\omega^2(\cos\omega t\,\mathbf{i} + \sin\omega t\,\mathbf{j}), a = \|\mathbf{a}\| = R\omega^2$, and $\mathbf{a} = -\omega^2\mathbf{r}$ is directed toward the origin.

(c) The smallest value of t for which $\mathbf{r}(t) = \mathbf{r}(0)$ satisfies $\omega t = 2\pi$, so $T = t = \dfrac{2\pi}{\omega}$.

33. **(a)** $\dfrac{d\mathbf{v}}{dt} = 2t^2\mathbf{i} + \mathbf{j} + \cos 2t\,\mathbf{k}, \mathbf{v}_0 = \mathbf{i} + 2\mathbf{j} - \mathbf{k}$, so $x'(t) = \dfrac{2}{3}t^3 + 1, y'(t) = t + 2, z'(t) = \dfrac{1}{2}\sin 2t - 1,$

$x(t) = \dfrac{1}{6}t^4 + t, y(t) = \dfrac{1}{2}t^2 + 2t, z(t) = -\dfrac{1}{4}\cos 2t - t + \dfrac{1}{4}$, since $\mathbf{r}(0) = \mathbf{0}$. Hence

$$\mathbf{r}(t) = \left(\dfrac{1}{6}t^4 + t\right)\mathbf{i} + \left(\dfrac{1}{2}t^2 + 2t\right)\mathbf{j} - \left(\dfrac{1}{4}\cos 2t + t - \dfrac{1}{4}\right)\mathbf{k}$$

(b) $\dfrac{ds}{dt}\Big]_{t=1} = \|\mathbf{r}'(t)\|\Big]_{t=1} \sqrt{(5/3)^2 + 9 + (1 - (\sin 2)/2)^2} \approx 3.475$

35. By equation (26) of Section 13.6, $\mathbf{r}(t) = (60\cos\alpha)t\,\mathbf{i} + ((60\sin\alpha)t - 16t^2 + 4)\mathbf{j}$, and the maximum height of the baseball occurs when $y'(t) = 0, 60\sin\alpha = 32t, t = \dfrac{15}{8}\sin\alpha$, so the ball clears the ceiling if $y_{\max} = (60\sin\alpha)\dfrac{15}{8}\sin\alpha - 16\dfrac{15^2}{8^2}\sin^2\alpha + 4 \le 25, \dfrac{15^2\sin^2\alpha}{4} \le 21, \sin^2\alpha \le \dfrac{28}{75}$. The ball hits the wall when $x = 60, t = \sec\alpha$, and $y(\sec\alpha) = 60\sin\alpha\sec\alpha - 16\sec^2\alpha + 4$. Maximize the height $h(\alpha) = y(\sec\alpha) = 60\tan\alpha - 16\sec^2\alpha + 4$, subject to the constraint $\sin^2\alpha \le \dfrac{28}{75}$. Then $h'(\alpha) = 60\sec^2\alpha - 32\sec^2\alpha\tan\alpha = 0, \tan\alpha = \dfrac{60}{32} = \dfrac{15}{8}$, so $\sin\alpha = \dfrac{15}{\sqrt{8^2 + 15^2}} = \dfrac{15}{17}$, but for this value of α the constraint is not satisfied (the ball hits the ceiling). Hence the maximum value of h occurs at one of the endpoints of the α-interval on which the ball clears the ceiling, i.e. $\left[0, \sin^{-1}\sqrt{28/75}\right]$. Since $h'(0) = 60$, it follows that h is increasing throughout the interval, since $h' > 0$ inside the interval. Thus $h_{\max}$ occurs when $\sin^2\alpha = \dfrac{28}{75}, h_{\max} = 60\tan\alpha - 16\sec^2\alpha + 4 =$

$$60\frac{\sqrt{28}}{\sqrt{47}} - 16\frac{75}{47} + 4 = \frac{120\sqrt{329} - 1012}{47} \approx 24.78 \text{ ft.}$$ Note: the possibility that the baseball keeps climbing until it hits the wall can be rejected as follows: if so, then $y'(t) = 0$ after the ball hits the wall, i.e. $t = \frac{15}{8}\sin\alpha$ occurs after $t = \sec\alpha$, hence $\frac{15}{8}\sin\alpha \geq \sec\alpha$, $15\sin\alpha\cos\alpha \geq 8$, $15\sin 2\alpha \geq 16$, impossible.

37. From Table 13.7.1, $GM \approx 3.99 \times 10^5 \text{ km}^3/\text{s}^2$, and $r_0 = 600$, so

$$v_{\text{esc}} = \sqrt{\frac{2GM}{r_0}} \approx \sqrt{\frac{3.99 \times 10^5}{300}} \approx 36.47 \text{ km/s}.$$

CHAPTER 14
Partial Derivatives

EXERCISE SET 14.1

1. **(a)** $f(2,1) = (2)^2(1) + 1 = 5$ **(b)** $f(1,2) = (1)^2(2) + 1 = 3$
   (c) $f(0,0) = (0)^2(0) + 1 = 1$ **(d)** $f(1,-3) = (1)^2(-3) + 1 = -2$
   (e) $f(3a,a) = (3a)^2(a) + 1 = 9a^3 + 1$ **(f)** $f(ab, a-b) = (ab)^2(a-b) + 1 = a^3b^2 - a^2b^3 + 1$

3. **(a)** $f(x+y, x-y) = (x+y)(x-y) + 3 = x^2 - y^2 + 3$
   (b) $f\left(xy, 3x^2y^3\right) = (xy)\left(3x^2y^3\right) + 3 = 3x^3y^4 + 3$

5. $F(g(x), h(y)) = F\left(x^3, 3y+1\right) = x^3 e^{x^3(3y+1)}$

7. **(a)** $t^2 + 3t^{10}$ **(b)** 0 **(c)** 3076

9. **(a)** $v = 7$ lies between $v = 5$ and $v = 15$, and $7 = 5 + 2 = 5 + \dfrac{2}{10}(15 - 5)$, so
$$WCI \approx 19 + \frac{2}{10}(13 - 19) = 19 - 1.2 = 17.8°\text{F}$$

   (b) $v = 28$ lies between $v = 25$ and $v = 30$, and $28 = 25 + \dfrac{3}{5}(30 - 25)$, so
$$WCI \approx 19 + \frac{3}{5}(25 - 19) = 19 + 3.6 = 22.6°\text{F}$$

11. **(a)** The depression is $20 - 16 = 4$, so the relative humidity is 66%.
    (b) The relative humidity $\approx 77 - (1/2)7 = 73.5\%$.
    (c) The relative humidity $\approx 59 + (2/5)4 = 60.6\%$.

13. **(a)** 19 **(b)** -9 **(c)** 3
    (d) $a^6 + 3$ **(e)** $-t^8 + 3$ **(f)** $(a+b)(a-b)^2b^3 + 3$

15. $F\left(x^2, y+1, z^2\right) = (y+1)e^{x^2(y+1)z^2}$

17. **(a)** $f(\sqrt{5}, 2, \pi, -3\pi) = 80\sqrt{\pi}$ **(b)** $f(1, 1, \ldots, 1) = \displaystyle\sum_{k=1}^{n} k = n(n+1)/2$

19.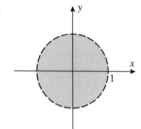

21.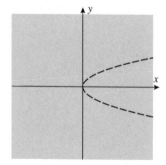

23. **(a)** all points above or on the line $y = -2$
    (b) all points on or within the sphere $x^2 + y^2 + z^2 = 25$
    (c) all points in 3-space

25.

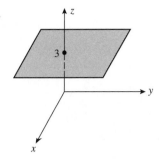

27.

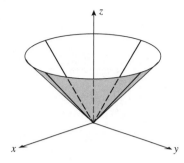

29.

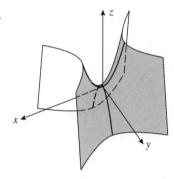

31.

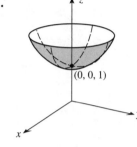

33.

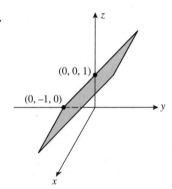

35. **(a)** $f(x,y) = 1 - x^2 - y^2$, because $f = c$ is a circle of radius $\sqrt{1-c}$ (provided $c \le 1$), and the radii in (a) decrease as c increases.

(b) $f(x,y) = \sqrt{x^2 + y^2}$ because $f = c$ is a circle of radius c, and the radii increase uniformly.

(c) $f(x,y) = x^2 + y^2$ because $f = c$ is a circle of radius $\sqrt{c}$ and the radii in the plot grow like the square root function.

37. **(a)** A **(b)** B **(c)** increase

(d) decrease   **(e)** increase   **(f)** decrease

39.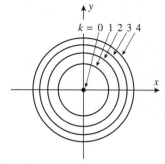

41.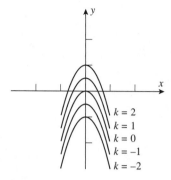

43.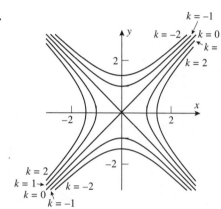

45.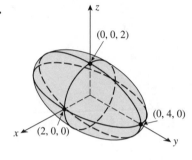

47.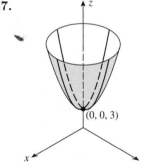

49. concentric spheres, common center at $(2,0,0)$

51. concentric cylinders, common axis the y-axis

53. (a) $f(-1,1) = 0$; $x^2 - 2x^3 + 3xy = 0$ (b) $f(0,0) = 0$; $x^2 - 2x^3 + 3xy = 0$
    (c) $f(2,-1) = -18$; $x^2 - 2x^3 + 3xy = -18$

55. (a) $f(1,-2,0) = 5$; $x^2 + y^2 - z = 5$ (b) $f(1,0,3) = -2$; $x^2 + y^2 - z = -2$
    (c) $f(0,0,0) = 0$; $x^2 + y^2 - z = 0$

57. (a)

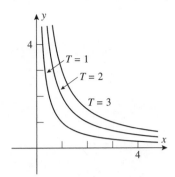

(b) At $(1, 4)$ the temperature is $T(1, 4) = 4$ so the temperature will remain constant along the path $xy = 4$.

59. (a)

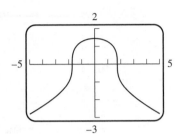

(b)

61. (a)

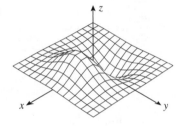

(b)

63. (a) The graph of g is the graph of f shifted one unit in the positive x-direction.

(b) The graph of g is the graph of f shifted one unit up the z-axis.

(c) The graph of g is the graph of f shifted one unit down the y-axis and then inverted with respect to the plane $z = 0$.

EXERCISE SET 14.2

1. 35   **3.** -8 **5.** 0

7. (a) Along $x = 0$ $\displaystyle\lim_{(x,y)\to(0,0)} \frac{3}{x^2 + 2y^2} = \lim_{y\to 0} \frac{3}{2y^2}$ does not exist.

(b) Along $x = 0$, $\displaystyle\lim_{(x,y)\to(0,0)} \frac{x + y}{2x^2 + y^2} = \lim_{y\to 0} \frac{1}{y}$ does not exist.

9. Let $z = x^2 + y^2$, then $\displaystyle\lim_{(x,y)\to(0,0)} \frac{\sin\left(x^2 + y^2\right)}{x^2 + y^2} = \lim_{z\to 0^+} \frac{\sin z}{z} = 1$

11. Let $z = x^2 + y^2$, then $\lim\limits_{(x,y)\to(0,0)} e^{-1/(x^2+y^2)} = \lim\limits_{z\to 0^+} e^{-1/z} = 0$

13. $\lim\limits_{(x,y)\to(0,0)} \dfrac{\left(x^2 + y^2\right)\left(x^2 - y^2\right)}{x^2 + y^2} = \lim\limits_{(x,y)\to(0,0)} \left(x^2 - y^2\right) = 0$

15. along $y = 0 : \lim\limits_{x\to 0} \dfrac{0}{3x^2} = \lim\limits_{x\to 0} 0 = 0$; along $y = x : \lim\limits_{x\to 0} \dfrac{x^2}{5x^2} = \lim\limits_{x\to 0} 1/5 = 1/5$
so the limit does not exist.

17. $8/3$

19. Let $t = \sqrt{x^2 + y^2 + z^2}$, then $\lim\limits_{(x,y,z)\to(0,0,0)} \dfrac{\sin\left(x^2 + y^2 + z^2\right)}{\sqrt{x^2 + y^2 + z^2}} = \lim\limits_{t\to 0^+} \dfrac{\sin\left(t^2\right)}{t} = 0$

21. $y\ln(x^2 + y^2) = r\sin\theta \ln r^2 = 2r(\ln r)\sin\theta$, so $\lim\limits_{(x,y)\to(0,0)} y\ln(x^2 + y^2) = \lim\limits_{r\to 0^+} 2r(\ln r)\sin\theta = 0$

23. $\dfrac{e^{\sqrt{x^2+y^2+z^2}}}{\sqrt{x^2 + y^2 + z^2}} = \dfrac{e^\rho}{\rho}$, so $\lim\limits_{(x,y,z)\to(0,0,0)} \dfrac{e^{\sqrt{x^2+y^2+z^2}}}{\sqrt{x^2 + y^2 + z^2}} = \lim\limits_{\rho\to 0^+} \dfrac{e^\rho}{\rho}$ does not exist.

25. (a) No, since there seem to be points near $(0,0)$ with $z = 0$ and other points near $(0,0)$ with $z \approx 1/2$.

(b) $\lim\limits_{x\to 0} \dfrac{mx^3}{x^4 + m^2 x^2} = \lim\limits_{x\to 0} \dfrac{mx}{x^2 + m^2} = 0$ **(c)** $\lim\limits_{x\to 0} \dfrac{x^4}{2x^4} = \lim\limits_{x\to 0} 1/2 = 1/2$

(d) A limit must be unique if it exists, so $f(x,y)$ cannot have a limit as $(x,y) \to (0,0)$.

27. (a) $\lim\limits_{t\to 0} \dfrac{abct^3}{a^2 t^2 + b^4 t^4 + c^4 t^4} = \lim\limits_{t\to 0} \dfrac{abct}{a^2 + b^4 t^2 + c^4 t^2} = 0$

(b) $\lim\limits_{t\to 0} \dfrac{t^4}{t^4 + t^4 + t^4} = \lim\limits_{t\to 0} 1/3 = 1/3$

29. $-\pi/2$ because $\dfrac{x^2 - 1}{x^2 + (y-1)^2} \to -\infty$ as $(x,y) \to (0,1)$

31. The required limit does not exist, so the singularity is not removeable.

33.

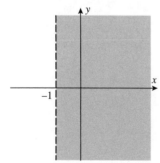

35.

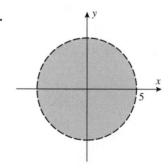

37.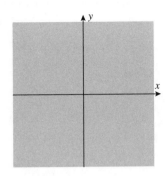

39.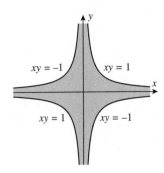

41. all of 3-space

43. all points not on the cylinder $x^2 + z^2 = 1$

EXERCISE SET 14.3

1. (a) $9x^2y^2$ (b) $6x^3y$ (c) $9y^2$ (d) $9x^2$
(e) $6y$ (f) $6x^3$ (g) 36 (h) 12

3. (a) $\dfrac{\partial z}{\partial x} = \dfrac{3}{2\sqrt{3x + 2y}}$; slope $= \dfrac{3}{8}$ (b) $\dfrac{\partial z}{\partial y} = \dfrac{1}{\sqrt{3x + 2y}}$; slope $= \dfrac{1}{4}$

5. (a) $\dfrac{\partial z}{\partial x} = -4\cos(y^2 - 4x)$; rate of change $= -4\cos 7$

(b) $\dfrac{\partial z}{\partial y} = 2y\cos(y^2 - 4x)$; rate of change $= 2\cos 7$

7. $\partial z/\partial x =$ slope of line parallel to xz-plane $= -4$; $\partial z/\partial y =$ slope of line parallel to yz-plane $= 1/2$

9. (a) The right-hand estimate is $\partial r/\partial v \approx (222 - 197)/(85 - 80) = 5$; the left-hand estimate is $\partial r/\partial v \approx (197 - 173)/(80 - 75) = 4.8$; the average is $\partial r/\partial v \approx 4.9$.

(b) The right-hand estimate is $\partial r/\partial \theta \approx (200 - 197)/(45 - 40) = 0.6$; the left-hand estimate is $\partial r/\partial \theta \approx (197 - 188)/(40 - 35) = 1.8$; the average is $\partial r/\partial \theta \approx 1.2$.

11. III is a plane, and its partial derivatives are constants, so III cannot be $f(x, y)$. If I is the graph of $z = f(x, y)$ then (by inspection) f_y is constant as y varies, but neither II nor III is constant as y varies. Hence $z = f(x, y)$ has II as its graph, and as II seems to be an odd function of x and an even function of y, f_x has I as its graph and f_y has III as its graph.

13. $\partial z/\partial x = 8xy^3 e^{x^2y^3}$, $\partial z/\partial y = 12x^2y^2 e^{x^2y^3}$

15. $\partial z/\partial x = x^3/(y^{3/5} + x) + 3x^2 \ln(1 + xy^{-3/5})$, $\partial z/\partial y = -(3/5)x^4/(y^{8/5} + xy)$

17. $\dfrac{\partial z}{\partial x} = -\dfrac{y(x^2 - y^2)}{(x^2 + y^2)^2}$, $\dfrac{\partial z}{\partial y} = \dfrac{x(x^2 - y^2)}{(x^2 + y^2)^2}$

19. $f_x(x, y) = (3/2)x^2 y \left(5x^2 - 7\right) \left(3x^5 y - 7x^3 y\right)^{-1/2}$
$f_y(x, y) = (1/2)x^3 \left(3x^2 - 7\right) \left(3x^5 y - 7x^3 y\right)^{-1/2}$

21. $f_x(x,y) = \dfrac{y^{-1/2}}{y^2 + x^2}$, $f_y(x,y) = -\dfrac{xy^{-3/2}}{y^2 + x^2} - \dfrac{3}{2}y^{-5/2}\tan^{-1}(x/y)$

23. $f_x(x,y) = -(4/3)y^2 \sec^2 x \left(y^2 \tan x\right)^{-7/3}$, $f_y(x,y) = -(8/3)y\tan x \left(y^2 \tan x\right)^{-7/3}$

25. $f_x(x,y) = -2x$, $f_x(3,1) = -6$; $f_y(x,y) = -21y^2$, $f_y(3,1) = -21$

27. $\partial z/\partial x = x(x^2 + 4y^2)^{-1/2}$, $\partial z/\partial x\,\big|_{(1,2)} = 1/\sqrt{17}$; $\partial z/\partial y = 4y(x^2 + 4y^2)^{-1/2}$, $\partial z/\partial y\,\big|_{(1,2)} = 8/\sqrt{17}$

29. (a) $2xy^4z^3 + y$ (b) $4x^2y^3z^3 + x$ (c) $3x^2y^4z^2 + 2z$
     (d) $2y^4z^3 + y$ (e) $32z^3 + 1$ (f) 438

31. $f_x = 2z/x$, $f_y = z/y$, $f_z = \ln(x^2 y \cos z) - z \tan z$

33. $f_x = -y^2z^3/\left(1 + x^2y^4z^6\right)$, $f_y = -2xyz^3/\left(1 + x^2y^4z^6\right)$, $f_z = -3xy^2z^2/\left(1 + x^2y^4z^6\right)$

35. $\partial w/\partial x = yze^z \cos xz$, $\partial w/\partial y = e^z \sin xz$, $\partial w/\partial z = ye^z(\sin xz + x\cos xz)$

37. $\partial w/\partial x = x/\sqrt{x^2 + y^2 + z^2}$, $\partial w/\partial y = y/\sqrt{x^2 + y^2 + z^2}$, $\partial w/\partial z = z/\sqrt{x^2 + y^2 + z^2}$

39. (a) e (b) $2e$ (c) e

41. (a) 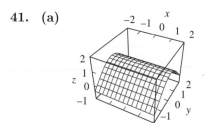   (b)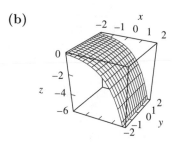

43. $\partial z/\partial x = 2x + 6y(\partial y/\partial x) = 2x$, $\partial z/\partial x\big]_{(2,1)} = 4$

45. $\partial z/\partial x = -x\left(29 - x^2 - y^2\right)^{-1/2}$, $\partial z/\partial x\big]_{(4,3)} = -2$

47. (a) $\partial V/\partial r = 2\pi rh$ (b) $\partial V/\partial h = \pi r^2$
     (c) $\partial V/\partial r]_{r=6,\,h=4} = 48\pi$ (d) $\partial V/\partial h]_{r=8,\,h=10} = 64\pi$

49. (a) $P = 10T/V$, $\partial P/\partial T = 10/V$, $\partial P/\partial T]_{T=80,\,V=50} = 1/5$ lb/(in^{2}K)
     (b) $V = 10T/P$, $\partial V/\partial P = -10T/P^2$, if $V = 50$ and $T = 80$ then
         $P = 10(80)/(50) = 16$, $\partial V/\partial P]_{T=80,\,P=16} = -25/8$(in^5/lb)

51. (a) $V = lwh$, $\partial V/\partial l = wh = 6$ (b) $\partial V/\partial w = lh = 15$
     (c) $\partial V/\partial h = lw = 10$

53. $\partial V/\partial r = \dfrac{2}{3}\pi rh = \dfrac{2}{r}(\dfrac{1}{3}\pi r^2 h) = 2V/r$

55. (a) $2x - 2z(\partial z/\partial x) = 0$, $\partial z/\partial x = x/z = \pm 3/(2\sqrt{6}) = \pm\sqrt{6}/4$
     (b) $z = \pm\sqrt{x^2 + y^2 - 1}$, $\partial z/\partial x = \pm x/\sqrt{x^2 + y^2 - 1} = \pm\sqrt{6}/4$

57. $\frac{3}{2}\left(x^2+y^2+z^2\right)^{1/2}\left(2x+2z\frac{\partial z}{\partial x}\right)=0$, $\partial z/\partial x=-x/z$; similarly, $\partial z/\partial y=-y/z$

59. $2x+z\left(xy\frac{\partial z}{\partial x}+yz\right)\cos xyz+\frac{\partial z}{\partial x}\sin xyz=0$, $\dfrac{\partial z}{\partial x}=-\dfrac{2x+yz^2\cos xyz}{xyz\cos xyz+\sin xyz}$;

$z\left(xy\frac{\partial z}{\partial y}+xz\right)\cos xyz+\frac{\partial z}{\partial y}\sin xyz=0$, $\dfrac{\partial z}{\partial y}=-\dfrac{xz^2\cos xyz}{xyz\cos xyz+\sin xyz}$

61. $(3/2)\left(x^2+y^2+z^2+w^2\right)^{1/2}\left(2x+2w\frac{\partial w}{\partial x}\right)=0$, $\partial w/\partial x=-x/w$; similarly, $\partial w/\partial y=-y/w$
and $\partial w/\partial z=-z/w$

63. $\dfrac{\partial w}{\partial x}=-\dfrac{yzw\cos xyz}{2w+\sin xyz}$, $\dfrac{\partial w}{\partial y}=-\dfrac{xzw\cos xyz}{2w+\sin xyz}$, $\dfrac{\partial w}{\partial z}=-\dfrac{xyw\cos xyz}{2w+\sin xyz}$

65. $f_x=e^{x^2}$, $f_y=-e^{y^2}$

67. (a) $-\dfrac{1}{4x^{3/2}}\cos y$ (b) $-\sqrt{x}\cos y$ (c) $-\dfrac{\sin y}{2\sqrt{x}}$ (d) $-\dfrac{\sin y}{2\sqrt{x}}$

69. $f_x=8x-8y^4$, $f_y=-32xy^3+35y^4$, $f_{xy}=f_{yx}=-32y^3$

71. $f_x=e^x\cos y$, $f_y=-e^x\sin y$, $f_{xy}=f_{yx}=-e^x\sin y$

73. $f_x=4/(4x-5y)$, $f_y=-5/(4x-5y)$, $f_{xy}=f_{yx}=20/(4x-5y)^2$

75. $f_x=2y/(x+y)^2$, $f_y=-2x/(x+y)^2$, $f_{xy}=f_{yx}=2(x-y)/(x+y)^3$

77. (a) $\dfrac{\partial^3 f}{\partial x^3}$ (b) $\dfrac{\partial^3 f}{\partial y^2\partial x}$ (c) $\dfrac{\partial^4 f}{\partial x^2\partial y^2}$ (d) $\dfrac{\partial^4 f}{\partial y^3\partial x}$

79. (a) $30xy^4-4$ (b) $60x^2y^3$ (c) $60x^3y^2$

81. (a) $f_{xyy}(0,1)=-30$ (b) $f_{xxx}(0,1)=-125$ (c) $f_{yyxx}(0,1)=150$

83. (a) $f_{xy}=15x^2y^4z^7+2y$ (b) $f_{yz}=35x^3y^4z^6+3y^2$
    (c) $f_{xz}=21x^2y^5z^6$ (d) $f_{zz}=42x^3y^5z^5$
    (e) $f_{zyy}=140x^3y^3z^6+6y$ (f) $f_{xxy}=30xy^4z^7$
    (g) $f_{zyx}=105x^2y^4z^6$ (h) $f_{xxyz}=210xy^4z^6$

85. (a) $f_x=2x+2y$, $f_{xx}=2$, $f_y=-2y+2x$, $f_{yy}=-2$; $f_{xx}+f_{yy}=2-2=0$
    (b) $z_x=e^x\sin y-e^y\sin x$, $z_{xx}=e^x\sin y-e^y\cos x$, $z_y=e^x\cos y+e^y\cos x$,
     $z_{yy}=-e^x\sin y+e^y\cos x$; $z_{xx}+z_{yy}=e^x\sin y-e^y\cos x-e^x\sin y+e^y\cos x=0$
    (c) $z_x=\dfrac{2x}{x^2+y^2}-2\dfrac{y}{x^2}\dfrac{1}{1+(y/x)^2}=\dfrac{2x-2y}{x^2+y^2}$, $z_{xx}=-2\dfrac{x^2-y^2-2xy}{(x^2+y^2)^2}$,

     $z_y=\dfrac{2y}{x^2+y^2}+2\dfrac{1}{x}\dfrac{1}{1+(y/x)^2}=\dfrac{2y+2x}{x^2+y^2}$, $z_{yy}=-2\dfrac{y^2-x^2+2xy}{(x^2+y^2)^2}$;

     $z_{xx}+z_{yy}=-2\dfrac{x^2-y^2-2xy}{(x^2+y^2)^2}-2\dfrac{y^2-x^2+2xy}{(x^2+y^2)^2}=0$

87. $u_x = \omega \sin c\omega t \cos \omega x$, $u_{xx} = -\omega^2 \sin c\omega t \sin \omega x$, $u_t = c\omega \cos c\omega t \sin \omega x$,

$u_{tt} = -c^2\omega^2 \sin c\omega t \sin \omega x$; $u_{xx} - \dfrac{1}{c^2}u_{tt} = -\omega^2 \sin c\omega t \sin \omega x - \dfrac{1}{c^2}(-c^2)\omega^2 \sin c\omega t \sin \omega x = 0$

89. $\partial u/\partial x = \partial v/\partial y$ and $\partial u/\partial y = -\partial v/\partial x$ so $\partial^2 u/\partial x^2 = \partial^2 v/\partial x \partial y$, and $\partial^2 u/\partial y^2 = -\partial^2 v/\partial y \partial x$, $\partial^2 u/\partial x^2 + \partial^2 u/\partial y^2 = \partial^2 v/\partial x \partial y - \partial^2 v/\partial y \partial x$, if $\partial^2 v/\partial x \partial y = \partial^2 v/\partial y \partial x$ then $\partial^2 u/\partial x^2 + \partial^2 u/\partial y^2 = 0$; thus u satisfies Laplace's equation. The proof that v satisfies Laplace's equation is similar. Adding Laplace's equations for u and v gives Laplaces' equation for $u + v$.

91. $\partial f/\partial v = 8vw^3 x^4 y^5$, $\partial f/\partial w = 12v^2 w^2 x^4 y^5$, $\partial f/\partial x = 16v^2 w^3 x^3 y^5$, $\partial f/\partial y = 20v^2 w^3 x^4 y^4$

93. $\partial f/\partial v_1 = 2v_1/\left(v_3^2 + v_4^2\right)$, $\partial f/\partial v_2 = -2v_2/\left(v_3^2 + v_4^2\right)$, $\partial f/\partial v_3 = -2v_3 \left(v_1^2 - v_2^2\right)/\left(v_3^2 + v_4^2\right)^2$, $\partial f/\partial v_4 = -2v_4 \left(v_1^2 - v_2^2\right)/\left(v_3^2 + v_4^2\right)^2$

95. **(a)** 0　　　　**(b)** 0　　　　**(c)** 0　　　　**(d)** 0

　　　(e) $2(1 + yw)e^{yw}\sin z \cos z$　　　　**(f)** $2xw(2 + yw)e^{yw}\sin z \cos z$

97. $\partial w/\partial x_i = -i\sin(x_1 + 2x_2 + \ldots + nx_n)$

99. **(a)** xy-plane, $f_x = 12x^2 y + 6xy$, $f_y = 4x^3 + 3x^2$, $f_{xy} = f_{yx} = 12x^2 + 6x$

　　　(b) $y \neq 0$, $f_x = 3x^2/y$, $f_y = -x^3/y^2$, $f_{xy} = f_{yx} = -3x^2/y^2$

101. $f_x(2, -1) = \lim\limits_{x \to 2} \dfrac{f(x, -1) - f(2, -1)}{x - 2} = \lim\limits_{x \to 2} \dfrac{2x^2 + 3x + 1 - 15}{x - 2} = \lim\limits_{x \to 2}(2x + 7) = 11$ and

$f_y(2, -1) = \lim\limits_{y \to -1} \dfrac{f(2, y) - f(2, -1)}{y + 1} = \lim\limits_{y \to -1} \dfrac{8 - 6y + y^2 - 15}{y + 1} = \lim\limits_{y \to -1} y - 7 = -8$

103. **(a)** $f_y(0, 0) = \dfrac{d}{dy}\big[f(0, y)\big]\bigg|_{y=0} = \dfrac{d}{dy}[y]\bigg|_{y=0} = 1$

　　　(b) If $(x, y) \neq (0, 0)$, then $f_y(x, y) = \dfrac{1}{3}(x^3 + y^3)^{-2/3}(3y^2) = \dfrac{y^2}{(x^3 + y^3)^{2/3}}$;

　　　　　$f_y(x, y)$ does not exist when $y \neq 0$ and $y = -x$

EXERCISE SET 14.4

1. $f(x, y) \approx f(3, 4) + f_x(x - 3) + f_y(y - 4) = 5 + 2(x - 3) - (y - 4)$ and
　　$f(3.01, 3.98) \approx 5 + 2(0.01) - (-0.02) = 5.04$

3. $L(x, y, z) = f(1, 2, 3) + (x - 1) + 2(y - 2) + 3(z - 3)$,
　　$f(1.01, 2.02, 3.03) \approx 4 + 0.01 + 2(0.02) + 3(0.03) = 4.14$

5. Suppose $f(x, y) = c$ for all (x, y). Then at (x_0, y_0) we have $\dfrac{f(x_0 + \Delta x, y_0) - f(x_0, y_0)}{\Delta x} = 0$ and hence $f_x(x_0, y_0)$ exists and is equal to 0 (Definition 14.3.1). A similar result holds for f_y. From equation (2), it follows that $\Delta f = 0$, and then by Definition 14.4.1 we see that f is differentiable at (x_0, y_0). An analogous result holds for functions $f(x, y, z)$ of three variables.

7. $f_x = 2x$, $f_y = 2y$, $f_z = 2z$ so $L(x, y, z) = 0$, $E = f - L = x^2 + y^2 + z^2$, and

$\lim\limits_{(x,y,z) \to (0,0,0)} \dfrac{E(x, y, z)}{\sqrt{x^2 + y^2 + z^2}} = \lim\limits_{(x,y,z) \to (0,0,0)} \sqrt{x^2 + y^2 + z^2} = 0$, so f is differentiable at $(0, 0, 0)$.

9. $dz = 7dx - 2dy$

11. $dz = 3x^2y^2dx + 2x^3ydy$

13. $dz = \left[y/\left(1 + x^2y^2\right)\right]dx + \left[x/\left(1 + x^2y^2\right)\right]dy$

15. $dw = 8dx - 3dy + 4dz$

17. $dw = 3x^2y^2zdx + 2x^3yzdy + x^3y^2dz$

19. $dw = \dfrac{yz}{1 + x^2y^2z^2}dx + \dfrac{xz}{1 + x^2y^2z^2}dy + \dfrac{xy}{1 + x^2y^2z^2}dz$

21. $df = (2x + 2y - 4)dx + 2xdy$; $x = 1$, $y = 2$, $dx = 0.01$, $dy = 0.04$ so
$df = 0.10$ and $\Delta f = 0.1009$

23. $df = -x^{-2}dx - y^{-2}dy$; $x = -1$, $y = -2$, $dx = -0.02$, $dy = -0.04$ so
$df = 0.03$ and $\Delta f \approx 0.029412$

25. $df = 2y^2z^3dx + 4xyz^3dy + 6xy^2z^2dz$, $x = 1, y = -1, z = 2, dx = -0.01, dy = -0.02, dz = 0.02$ so
$df = 0.96$ and $\Delta f \approx 0.97929$

27. Label the four smaller rectangles A, B, C, D starting with the lower left and going clockwise. Then the increase in the area of the rectangle is represented by B, C and D; and the portions B and D represent the approximation of the increase in area given by the total differential.

29. **(a)** $f(P) = 1/5, f_x(P) = -x/(x^2 + y^2)^{-3/2}\Big]_{(x,y)=(4,3)} = -4/125,$

$f_y(P) = -y/(x^2 + y^2)^{-3/2}\Big]_{(x,y)=(4,3)} = -3/125, L(x, y) = \dfrac{1}{5} - \dfrac{4}{125}(x - 4) - \dfrac{3}{125}(y - 3)$

(b) $L(Q) - f(Q) = \dfrac{1}{5} - \dfrac{4}{125}(-0.08) - \dfrac{3}{125}(0.01) - 0.2023342382 \approx -0.0000142382,$

$|PQ| = \sqrt{0.08^2 + 0.01^2} \approx 0.0008062257748, |L(Q) - f(Q)|/|PQ| \approx 0.000176603$

31. **(a)** $f(P) = 0, f_x(P) = 0, f_y(P) = 0, L(x, y) = 0$

(b) $L(Q) - f(Q) = -0.003\sin(0.004) \approx -0.000012, |PQ| = \sqrt{0.003^2 + 0.004^2} = 0.005,$
$|L(Q) - f(Q)|/|PQ| \approx 0.0024$

33. **(a)** $f(P) = 6, f_x(P) = 6, f_y(P) = 3, f_z(P) = 2, L(x, y) = 6 + 6(x - 1) + 3(y - 2) + 2(z - 3)$

(b) $L(Q) - f(Q) = 6 + 6(0.001) + 3(0.002) + 2(0.003) - 6.018018006 = -.000018006,$
$|PQ| = \sqrt{0.001^2 + 0.002^2 + 0.003^2} \approx .0003741657387; \quad |L(Q) - f(Q)|/|PQ| \approx -0.000481$

35. **(a)** $f(P) = e, f_x(P) = e, f_y(P) = -e, f_z(P) = -e, L(x, y) = e + e(x - 1) - e(y + 1) - e(z + 1)$

(b) $L(Q) - f(Q) = e - 0.01e + 0.01e - 0.01e - 0.99e^{0.9999} = 0.99(e - e^{0.9999}),$
$|PQ| = \sqrt{0.01^2 + 0.01^2 + 0.01^2} \approx 0.01732, |L(Q) - f(Q)|/|PQ| \approx 0.01554$

37. **(a)** Let $f(x, y) = e^x\sin y$; $f(0, 0) = 0, f_x(0, 0) = 0, f_y(0, 0) = 1$, so $e^x\sin y \approx y$

(b) Let $f(x, y) = \dfrac{2x + 1}{y + 1}$; $f(0, 0) = 1, f_x(0, 0) = 2, f_y(0, 0) = -1$, so $\dfrac{2x + 1}{y + 1} \approx 1 + 2x - y$

39. **(a)** Let $f(x, y, z) = xyz + 2$, then $f_x = f_y = f_z = 1$ at $x = y = z = 1$, and
$L(x, y, z) = f(1, 1, 1) + f_x(x - 1) + f_y(y - 1) + f_z(z - 1) = 3 + x - 1 + y - 1 + z - 1 = x + y + z$

(b) Let $f(x,y,z) = \dfrac{4x}{y+z}$, then $f_x = 2, f_y = -1, f_z = -1$ at $x = y = z = 1$, and

$$L(x,y,z) = f(1,1,1) + f_x(x-1) + f_y(y-1) + f_z(z-1)$$
$$= 2 + 2(x-1) - (y-1) - (z-1) = 2x - y - z + 2$$

41. $L(x,y) = f(1,1) + f_x(1,1)(x-1) + f_y(1,1)(y-1)$ and
$L(1.1, 0.9) = 3.15 = 3 + 2(0.1) + f_y(1,1)(-0.1)$ so $f_y(1,1) = -0.05/(-0.1) = 0.5$

43. $x - y + 2z - 2 = L(x,y,z) = f(3,2,1) + f_x(3,2,1)(x-3) + f_y(3,2,1)(y-2) + f_z(3,2,1)(z-1)$, so
$f_x(3,2,1) = 1, f_y(3,2,1) = -1, f_z(3,2,1) = 2$ and $f(3,2,1) = L(3,2,1) = 1$

45. $L(x,y) = f(x_0, y_0) + f_x(x_0, y_0)(x - x_0) + f_y(x_0, y_0)(y - y_0),$
$2y - 2x - 2 = x_0^2 + y_0^2 + 2x_0(x - x_0) + 2y_0(y - y_0)$, from which it follows that $x_0 = -1, y_0 = 1$.

47. $L(x,y,z) = f(x_0, y_0, z_0) + f_x(x_0, y_0, z_0)(x - x_0) + f_y(x_0, y_0, z_0)(y - y_0) + f_z(x_0, y_0, z_0)(z - z_0),$
$y + 2z - 1 = x_0 y_0 + z_0^2 + y_0(x - x_0) + x_0(y - y_0) + 2z_0(z - z_0)$, so that $x_0 = 1, y_0 = 0, z_0 = 1$.

49. $A = xy, dA = y\,dx + x\,dy, dA/A = dx/x + dy/y, |dx/x| \le 0.03$ and $|dy/y| \le 0.05,$
$|dA/A| \le |dx/x| + |dy/y| \le 0.08 = 8\%$

51. $z = \sqrt{x^2 + y^2}, dz = \dfrac{x}{\sqrt{x^2 + y^2}}dx + \dfrac{y}{\sqrt{x^2 + y^2}}dy,$

$\dfrac{dz}{z} = \dfrac{x}{x^2 + y^2}dx + \dfrac{y}{x^2 + y^2}dy = \dfrac{x^2}{x^2 + y^2}\left(\dfrac{dx}{x}\right) + \dfrac{y^2}{x^2 + y^2}\left(\dfrac{dy}{y}\right),$

$\left|\dfrac{dz}{z}\right| \le \dfrac{x^2}{x^2 + y^2}\left|\dfrac{dx}{x}\right| + \dfrac{y^2}{x^2 + y^2}\left|\dfrac{dy}{y}\right|$, if $\left|\dfrac{dx}{x}\right| \le r/100$ and $\left|\dfrac{dy}{y}\right| \le r/100$ then

$\left|\dfrac{dz}{z}\right| \le \dfrac{x^2}{x^2 + y^2}(r/100) + \dfrac{y^2}{x^2 + y^2}(r/100) = \dfrac{r}{100}$ so the percentage error in z is at most about $r\%$.

53. $dT = \dfrac{\pi}{g\sqrt{L/g}}dL - \dfrac{\pi L}{g^2 \sqrt{L/g}}dg, \dfrac{dT}{T} = \dfrac{1}{2}\dfrac{dL}{L} - \dfrac{1}{2}\dfrac{dg}{g}; |dL/L| \le 0.005$ and $|dg/g| \le 0.001$ so
$|dT/T| \le (1/2)(0.005) + (1/2)(0.001) = 0.003 = 0.3\%$

55. (a) $\left|\dfrac{d(xy)}{xy}\right| = \left|\dfrac{y\,dx + x\,dy}{xy}\right| = \left|\dfrac{dx}{x} + \dfrac{dy}{y}\right| \le \left|\dfrac{dx}{x}\right| + \left|\dfrac{dy}{y}\right| \le \dfrac{r}{100} + \dfrac{s}{100}; (r+s)\%$

(b) $\left|\dfrac{d(x/y)}{x/y}\right| = \left|\dfrac{y\,dx - x\,dy}{xy}\right| = \left|\dfrac{dx}{x} - \dfrac{dy}{y}\right| \le \left|\dfrac{dx}{x}\right| + \left|\dfrac{dy}{y}\right| \le \dfrac{r}{100} + \dfrac{s}{100}; (r+s)\%$

(c) $\left|\dfrac{d(x^2 y^3)}{x^2 y^3}\right| = \left|\dfrac{2xy^3\,dx + 3x^2 y^2\,dy}{x^2 y^3}\right| = \left|2\dfrac{dx}{x} + 3\dfrac{dy}{y}\right| \le 2\left|\dfrac{dx}{x}\right| + 3\left|\dfrac{dy}{y}\right|$
$\le 2\dfrac{r}{100} + 3\dfrac{s}{100}; (2r + 3s)\%$

(d) $\left|\dfrac{d(x^3 y^{1/2})}{x^3 y^{1/2}}\right| = \left|\dfrac{3x^2 y^{1/2}\,dx + (1/2)x^3 y^{-1/2}\,dy}{x^3 y^{1/2}}\right| = \left|3\dfrac{dx}{x} + \dfrac{1}{2}\dfrac{dy}{y}\right| \le 3\left|\dfrac{dx}{x}\right| + \dfrac{1}{2}\left|\dfrac{dy}{y}\right|$
$\le 3\dfrac{r}{100} + \dfrac{1}{2}\dfrac{s}{100}; (3r + \tfrac{1}{2}s)\%$

57. $dA = \dfrac{1}{2}b\sin\theta\, da + \dfrac{1}{2}a\sin\theta\, db + \dfrac{1}{2}ab\cos\theta\, d\theta,$

$|dA| \le \dfrac{1}{2}b\sin\theta|da| + \dfrac{1}{2}a\sin\theta|db| + \dfrac{1}{2}ab\cos\theta|d\theta|$

$\le \dfrac{1}{2}(50)(1/2)(1/2) + \dfrac{1}{2}(40)(1/2)(1/4) + \dfrac{1}{2}(40)(50)\left(\sqrt{3}/2\right)(\pi/90)$

$= 35/4 + 50\pi\sqrt{3}/9 \approx 39 \text{ ft}^2$

59. $f_x = 2x\sin y$, $f_y = x^2\cos y$ are both continuous everywhere, so f is differentiable everywhere.

61. That f is differentiable means that $\displaystyle\lim_{(x,y)\to(x_0,y_0)} \dfrac{E_f(x,y)}{\sqrt{(x-x_0)^2 + (y-y_0)^2}} = 0$, where
$E_f(x,y) = f(x,y) - L_f(x,y)$; here $L_f(x,y)$ is the linear approximation to f at (x_0, y_0).
Let f_x and f_y denote $f_x(x_0, y_0)$, $f_y(x_0, y_0)$ respectively. Then $g(x,y,z) = z - f(x,y)$,
$L_f(x,y) = f(x_0, y_0) + f_x(x - x_0) + f_y(y - y_0)$,
$L_g(x,y,z) = g(x_0, y_0, z_0) + g_x(x - x_0) + g_y(y - y_0) + g_z(z - z_0)$,
$= 0 - f_x(x - x_0) - f_y(y - y_0) + (z - z_0)$
and

$E_g(x,y,z) = g(x,y,z) - L_g(x,y,z) = (z - f(x,y)) + f_x(x - x_0) + f_y(y - y_0) - (z - z_0)$
$= f(x_0, y_0) + f_x(x_0, y_0)(x - x_0) + f_y(x_0, y_0)(y - y_0) - f(x,y) = -E_f(x,y)$

Thus $\dfrac{|E_g(x,y,z)|}{\sqrt{(x-x_0)^2 + (y-y_0)^2 + (z-z_0)^2}} \le \dfrac{|E_f(x,y)|}{\sqrt{(x-x_0)^2 + (y-y_0)^2}}$

so $\displaystyle\lim_{(x,y,z)\to(x_0,y_0,z_0)} \dfrac{E_g(x,y,z)}{\sqrt{(x-x_0)^2 + (y-y_0)^2 + (z-z_0)^2}} = 0$

and g is differentiable at (x_0, y_0, z_0).

EXERCISE SET 14.5

1. $42t^{13}$

3. $3t^{-2}\sin(1/t)$

5. $-\dfrac{10}{3}t^{7/3}e^{1-t^{10/3}}$

7. $165t^{32}$

9. $-2t\cos\left(t^2\right)$

11. 3264

13. $\dfrac{dz}{dt} = \dfrac{\partial z}{\partial x}\dfrac{dx}{dt} + \dfrac{\partial z}{\partial y}\dfrac{dy}{dt} = 3(2t)_{t=2} - (3t^2)_{t=2} = 12 - 12 = 0$

17. $\partial z/\partial u = 24u^2 v^2 - 16uv^3 - 2v + 3$, $\partial z/\partial v = 16u^3 v - 24u^2 v^2 - 2u - 3$

19. $\partial z/\partial u = -\dfrac{2\sin u}{3\sin v}$, $\partial z/\partial v = -\dfrac{2\cos u\cos v}{3\sin^2 v}$

21. $\partial z/\partial u = e^u$, $\partial z/\partial v = 0$

23. $\partial T/\partial r = 3r^2\sin\theta\cos^2\theta - 4r^3\sin^3\theta\cos\theta$
$\partial T/\partial\theta = -2r^3\sin^2\theta\cos\theta + r^4\sin^4\theta + r^3\cos^3\theta - 3r^4\sin^2\theta\cos^2\theta$

25. $\partial t/\partial x = \left(x^2 + y^2\right)/\left(4x^2 y^3\right)$, $\partial t/\partial y = \left(y^2 - 3x^2\right)/\left(4xy^4\right)$

27. $\partial z/\partial r = (dz/dx)(\partial x/\partial r) = 2r\cos^2\theta/\left(r^2\cos^2\theta + 1\right),$

$\partial z/\partial\theta = (dz/dx)(\partial x/\partial\theta) = -2r^2\sin\theta\cos\theta/\left(r^2\cos^2\theta + 1\right)$

29. $\partial w/\partial\rho = 2\rho\left(4\sin^2\phi + \cos^2\phi\right),\ \partial w/\partial\phi = 6\rho^2\sin\phi\cos\phi,\ \partial w/\partial\theta = 0$

31. $-\pi$
 33. $\sqrt{3}e^{\sqrt{3}},\ \left(2 - 4\sqrt{3}\right)e^{\sqrt{3}}$

35. $F(x,y) = x^2y^3 + \cos y,\ \dfrac{dy}{dx} = -\dfrac{2xy^3}{3x^2y^2 - \sin y}$

37. $F(x,y) = e^{xy} + ye^y - 1,\ \dfrac{dy}{dx} = -\dfrac{ye^{xy}}{xe^{xy} + ye^y + e^y}$

39. $\dfrac{\partial F}{\partial x} + \dfrac{\partial F}{\partial z}\dfrac{\partial z}{\partial x} = 0$ so $\dfrac{\partial z}{\partial x} = -\dfrac{\partial F/\partial x}{\partial F/\partial z}.$
 41. $\dfrac{\partial z}{\partial x} = \dfrac{2x + yz}{6yz - xy},\ \dfrac{\partial z}{\partial y} = \dfrac{xz - 3z^2}{6yz - xy}$

43. $ye^x - 5\sin 3z - 3z = 0;\ \dfrac{\partial z}{\partial x} = -\dfrac{ye^x}{-15\cos 3z - 3} = \dfrac{ye^x}{15\cos 3z + 3},\ \dfrac{\partial z}{\partial y} = \dfrac{e^x}{15\cos 3z + 3}$

45. $D = \left(x^2 + y^2\right)^{1/2}$ where x and y are the distances of cars A and B, respectively, from the intersection and D is the distance between them.

$dD/dt = \left[x/\left(x^2 + y^2\right)^{1/2}\right](dx/dt) + \left[y/\left(x^2 + y^2\right)^{1/2}\right](dy/dt),\ dx/dt = -25$ and $dy/dt = -30$ when $x = 0.3$ and $y = 0.4$ so $dD/dt = (0.3/0.5)(-25) + (0.4/0.5)(-30) = -39$ mph.

47. $A = \dfrac{1}{2}ab\sin\theta$ but $\theta = \pi/6$ when $a = 4$ and $b = 3$ so $A = \dfrac{1}{2}(4)(3)\sin(\pi/6) = 3.$

Solve $\dfrac{1}{2}ab\sin\theta = 3$ for θ to get $\theta = \sin^{-1}\left(\dfrac{6}{ab}\right),\ 0 \le \theta \le \pi/2.$

$\dfrac{d\theta}{dt} = \dfrac{\partial\theta}{\partial a}\dfrac{da}{dt} + \dfrac{\partial\theta}{\partial b}\dfrac{db}{dt} = \dfrac{1}{\sqrt{1 - \dfrac{36}{a^2b^2}}}\left(-\dfrac{6}{a^2b}\right)\dfrac{da}{dt} + \dfrac{1}{\sqrt{1 - \dfrac{36}{a^2b^2}}}\left(-\dfrac{6}{ab^2}\right)\dfrac{db}{dt}$

$= -\dfrac{6}{\sqrt{a^2b^2 - 36}}\left(\dfrac{1}{a}\dfrac{da}{dt} + \dfrac{1}{b}\dfrac{db}{dt}\right),\ \dfrac{da}{dt} = 1$ and $\dfrac{db}{dt} = 1$

when $a = 4$ and $b = 3$ so $\dfrac{d\theta}{dt} = -\dfrac{6}{\sqrt{144 - 36}}\left(\dfrac{1}{4} + \dfrac{1}{3}\right) = -\dfrac{7}{12\sqrt{3}} = -\dfrac{7}{36}\sqrt{3}$ radians/s

49. $V = (\pi/4)D^2h$ where D is the diameter and h is the height, both measured in inches, $dV/dt = (\pi/2)Dh(dD/dt) + (\pi/4)D^2(dh/dt),\ dD/dt = 3$ and $dh/dt = 24$ when $D = 30$ and $h = 240$, so $dV/dt = (\pi/2)(30)(240)(3) + (\pi/4)(30)^2(24) = 16{,}200\pi$ in^3/year.

51. **(a)** $V = \ell wh,\ \dfrac{dV}{dt} = \dfrac{\partial V}{\partial\ell}\dfrac{d\ell}{dt} + \dfrac{\partial V}{\partial w}\dfrac{dw}{dt} + \dfrac{\partial V}{\partial h}\dfrac{dh}{dt} = wh\dfrac{d\ell}{dt} + \ell h\dfrac{dw}{dt} + \ell w\dfrac{dh}{dt}$

$= (3)(6)(1) + (2)(6)(2) + (2)(3)(3) = 60$ in^3/s

(b) $D = \sqrt{\ell^2 + w^2 + h^2};\ dD/dt = (\ell/D)d\ell/dt + (w/D)dw/dt + (h/D)dh/dt$

$= (2/7)(1) + (3/7)(2) + (6/7)(3) = 26/7$ in/s

53. **(a)** $f(tx, ty) = 3t^2x^2 + t^2y^2 = t^2 f(x, y);\ n = 2$

(b) $f(tx, ty) = \sqrt{t^2x^2 + t^2y^2} = tf(x, y);\ n = 1$

(c) $f(tx, ty) = t^3x^2y - 2t^3y^3 = t^3 f(x, y);\ n = 3$

(d) $f(tx, ty) = 5/\left(t^2x^2 + 2t^2y^2\right)^2 = t^{-4} f(x, y);\ n = -4$

55. **(a)** $\dfrac{\partial z}{\partial x} = \dfrac{dz}{du}\dfrac{\partial u}{\partial x},\ \dfrac{\partial z}{\partial y} = \dfrac{dz}{du}\dfrac{\partial u}{\partial y}$

(b) $\dfrac{\partial^2 z}{\partial x^2} = \dfrac{dz}{du}\dfrac{\partial^2 u}{\partial x^2} + \dfrac{\partial}{\partial x}\left(\dfrac{dz}{du}\right)\dfrac{\partial u}{\partial x} = \dfrac{dz}{du}\dfrac{\partial^2 u}{\partial x^2} + \dfrac{d^2 z}{du^2}\left(\dfrac{\partial u}{\partial x}\right)^2;$

$\dfrac{\partial^2 z}{\partial y \partial x} = \dfrac{dz}{du}\dfrac{\partial^2 u}{\partial y \partial x} + \dfrac{\partial}{\partial y}\left(\dfrac{dz}{du}\right)\dfrac{\partial u}{\partial x} = \dfrac{dz}{du}\dfrac{\partial^2 u}{\partial y \partial x} + \dfrac{d^2 z}{du^2}\dfrac{\partial u}{\partial x}\dfrac{\partial u}{\partial y}$

$\dfrac{\partial^2 z}{\partial y^2} = \dfrac{dz}{du}\dfrac{\partial^2 u}{\partial y^2} + \dfrac{\partial}{\partial y}\left(\dfrac{dz}{du}\right)\dfrac{\partial u}{\partial y} = \dfrac{dz}{du}\dfrac{\partial^2 u}{\partial y^2} + \dfrac{d^2 z}{du^2}\left(\dfrac{\partial u}{\partial y}\right)^2$

57. Let $z = f(u)$ where $u = x + 2y$; then $\partial z/\partial x = (dz/du)(\partial u/\partial x) = dz/du$,

$\partial z/\partial y = (dz/du)(\partial u/\partial y) = 2dz/du$ so $2\partial z/\partial x - \partial z/\partial y = 2dz/du - 2dz/du = 0$

59. $\dfrac{\partial w}{\partial x} = \dfrac{dw}{du}\dfrac{\partial u}{\partial x} = \dfrac{dw}{du},\ \dfrac{\partial w}{\partial y} = \dfrac{dw}{du}\dfrac{\partial u}{\partial y} = 2\dfrac{dw}{du},\ \dfrac{\partial w}{\partial z} = \dfrac{dw}{du}\dfrac{\partial u}{\partial z} = 3\dfrac{dw}{du},$ so $\dfrac{\partial w}{\partial x} + \dfrac{\partial w}{\partial y} + \dfrac{\partial w}{\partial z} = 6\dfrac{dw}{du}$

61. $z = f(u, v)$ where $u = x - y$ and $v = y - x$,

$\dfrac{\partial z}{\partial x} = \dfrac{\partial z}{\partial u}\dfrac{\partial u}{\partial x} + \dfrac{\partial z}{\partial v}\dfrac{\partial v}{\partial x} = \dfrac{\partial z}{\partial u} - \dfrac{\partial z}{\partial v}$ and $\dfrac{\partial z}{\partial y} = \dfrac{\partial z}{\partial u}\dfrac{\partial u}{\partial y} + \dfrac{\partial z}{\partial v}\dfrac{\partial v}{\partial y} = -\dfrac{\partial z}{\partial u} + \dfrac{\partial z}{\partial v}$ so $\dfrac{\partial z}{\partial x} + \dfrac{\partial z}{\partial y} = 0$

63. **(a)** $1 = -r\sin\theta\,\dfrac{\partial\theta}{\partial x} + \cos\theta\,\dfrac{\partial r}{\partial x}$ and $0 = r\cos\theta\,\dfrac{\partial\theta}{\partial x} + \sin\theta\,\dfrac{\partial r}{\partial x}$; solve for $\partial r/\partial x$ and $\partial\theta/\partial x$.

(b) $0 = -r\sin\theta\,\dfrac{\partial\theta}{\partial y} + \cos\theta\,\dfrac{\partial r}{\partial y}$ and $1 = r\cos\theta\,\dfrac{\partial\theta}{\partial y} + \sin\theta\,\dfrac{\partial r}{\partial y}$; solve for $\partial r/\partial y$ and $\partial\theta/\partial y$.

(c) $\dfrac{\partial z}{\partial x} = \dfrac{\partial z}{\partial r}\dfrac{\partial r}{\partial x} + \dfrac{\partial z}{\partial\theta}\dfrac{\partial\theta}{\partial x} = \dfrac{\partial z}{\partial r}\cos\theta - \dfrac{1}{r}\dfrac{\partial z}{\partial\theta}\sin\theta.$

$\dfrac{\partial z}{\partial y} = \dfrac{\partial z}{\partial r}\dfrac{\partial r}{\partial y} + \dfrac{\partial z}{\partial\theta}\dfrac{\partial\theta}{\partial y} = \dfrac{\partial z}{\partial r}\sin\theta + \dfrac{1}{r}\dfrac{\partial z}{\partial\theta}\cos\theta.$

(d) Square and add the results of Parts (a) and (b).

(e) From Part (c),

$\dfrac{\partial^2 z}{\partial x^2} = \dfrac{\partial}{\partial r}\left(\dfrac{\partial z}{\partial r}\cos\theta - \dfrac{1}{r}\dfrac{\partial z}{\partial\theta}\sin\theta\right)\dfrac{\partial r}{\partial x} + \dfrac{\partial}{\partial\theta}\left(\dfrac{\partial z}{\partial r}\cos\theta - \dfrac{1}{r}\dfrac{\partial z}{\partial\theta}\sin\theta\right)\dfrac{\partial\theta}{\partial x}$

$= \left(\dfrac{\partial^2 z}{\partial r^2}\cos\theta + \dfrac{1}{r^2}\dfrac{\partial z}{\partial\theta}\sin\theta - \dfrac{1}{r}\dfrac{\partial^2 z}{\partial r\partial\theta}\sin\theta\right)\cos\theta$

$+ \left(\dfrac{\partial^2 z}{\partial\theta\partial r}\cos\theta - \dfrac{\partial z}{\partial r}\sin\theta - \dfrac{1}{r}\dfrac{\partial^2 z}{\partial\theta^2}\sin\theta - \dfrac{1}{r}\dfrac{\partial z}{\partial\theta}\cos\theta\right)\left(-\dfrac{\sin\theta}{r}\right)$

$= \dfrac{\partial^2 z}{\partial r^2}\cos^2\theta + \dfrac{2}{r^2}\dfrac{\partial z}{\partial\theta}\sin\theta\cos\theta - \dfrac{2}{r}\dfrac{\partial^2 z}{\partial\theta\partial r}\sin\theta\cos\theta + \dfrac{1}{r^2}\dfrac{\partial^2 z}{\partial\theta^2}\sin^2\theta + \dfrac{1}{r}\dfrac{\partial z}{\partial r}\sin^2\theta.$

Similarly, from Part (c),

$$\frac{\partial^2 z}{\partial y^2} = \frac{\partial^2 z}{\partial r^2}\sin^2\theta - \frac{2}{r^2}\frac{\partial z}{\partial\theta}\sin\theta\cos\theta + \frac{2}{r}\frac{\partial^2 z}{\partial\theta\partial r}\sin\theta\cos\theta + \frac{1}{r^2}\frac{\partial^2 z}{\partial\theta^2}\cos^2\theta + \frac{1}{r}\frac{\partial z}{\partial r}\cos^2\theta.$$

Add to get $\dfrac{\partial^2 z}{\partial x^2} + \dfrac{\partial^2 z}{\partial y^2} = \dfrac{\partial^2 z}{\partial r^2} + \dfrac{1}{r^2}\dfrac{\partial^2 z}{\partial\theta^2} + \dfrac{1}{r}\dfrac{\partial z}{\partial r}$.

65. **(a)** By the chain rule, $\dfrac{\partial u}{\partial r} = \dfrac{\partial u}{\partial x}\cos\theta + \dfrac{\partial u}{\partial y}\sin\theta$ and $\dfrac{\partial v}{\partial\theta} = -\dfrac{\partial v}{\partial x}r\sin\theta + \dfrac{\partial v}{\partial y}r\cos\theta$, use the

Cauchy-Riemann conditions $\dfrac{\partial u}{\partial x} = \dfrac{\partial v}{\partial y}$ and $\dfrac{\partial u}{\partial y} = -\dfrac{\partial v}{\partial x}$ in the equation for $\dfrac{\partial u}{\partial r}$ to get

$\dfrac{\partial u}{\partial r} = \dfrac{\partial v}{\partial y}\cos\theta - \dfrac{\partial v}{\partial x}\sin\theta$ and compare to $\dfrac{\partial v}{\partial\theta}$ to see that $\dfrac{\partial u}{\partial r} = \dfrac{1}{r}\dfrac{\partial v}{\partial\theta}$. The result $\dfrac{\partial v}{\partial r} = -\dfrac{1}{r}\dfrac{\partial u}{\partial\theta}$

can be obtained by considering $\dfrac{\partial v}{\partial r}$ and $\dfrac{\partial u}{\partial\theta}$.

(b) $u_x = \dfrac{2x}{x^2+y^2}$, $v_y = 2\dfrac{1}{x}\dfrac{1}{1+(y/x)^2} = \dfrac{2x}{x^2+y^2} = u_x$;

$u_y = \dfrac{2y}{x^2+y^2}$, $v_x = -2\dfrac{y}{x^2}\dfrac{1}{1+(y/x)^2} = -\dfrac{2y}{x^2+y^2} = -u_y$;

$u = \ln r^2$, $v = 2\theta$, $u_r = 2/r$, $v_\theta = 2$, so $u_r = \dfrac{1}{r}v_\theta$, $u_\theta = 0$, $v_r = 0$, so $v_r = -\dfrac{1}{r}u_\theta$

67. $\partial w/\partial\rho = (\sin\phi\cos\theta)\partial w/\partial x + (\sin\phi\sin\theta)\partial w/\partial y + (\cos\phi)\,\partial w/\partial z$
$\partial w/\partial\phi = (\rho\cos\phi\cos\theta)\partial w/\partial x + (\rho\cos\phi\sin\theta)\partial w/\partial y - (\rho\sin\phi)\partial w/\partial z$
$\partial w/\partial\theta = -(\rho\sin\phi\sin\theta)\partial w/\partial x + (\rho\sin\phi\cos\theta)\partial w/\partial y$

69. $w_r = e^r/(e^r + e^s + e^t + e^u)$, $w_{rs} = -e^r e^s/(e^r + e^s + e^t + e^u)^2$,
$w_{rst} = 2e^r e^s e^t/(e^r + e^s + e^t + e^u)^3$,
$w_{rstu} = -6e^r e^s e^t e^u/(e^r + e^s + e^t + e^u)^4 = -6e^{r+s+t+u}/e^{4w} = -6e^{r+s+t+u-4w}$

71. **(a)** $dw/dt = \displaystyle\sum_{i=1}^{4}(\partial w/\partial x_i)\,(dx_i/dt)$

(b) $\partial w/\partial v_j = \displaystyle\sum_{i=1}^{4}(\partial w/\partial x_i)\,(\partial x_i/\partial v_j)$ for $j = 1, 2, 3$

73. $dF/dx = (\partial F/\partial u)(du/dx) + (\partial F/\partial v)(dv/dx)$
$\quad\quad = f(u)g'(x) - f(v)h'(x) = f(g(x))g'(x) - f(h(x))h'(x)$

75. Let (a, b) be any point in the region, if (x, y) is in the region then by the result of Exercise 74
$f(x, y) - f(a, b) = f_x(x^*, y^*)(x - a) + f_y(x^*, y^*)(y - b)$ where (x^*, y^*) is on the line segment joining
(a, b) and (x, y). If $f_x(x, y) = f_y(x, y) = 0$ throughout the region then
$f(x, y) - f(a, b) = (0)(x - a) + (0)(y - b) = 0$, $f(x, y) = f(a, b)$ so $f(x, y)$ is constant on the region.

EXERCISE SET 14.6

1. $\nabla f(x, y) = (3y/2)(1 + xy)^{1/2}\mathbf{i} + (3x/2)(1 + xy)^{1/2}\mathbf{j}$, $\nabla f(3, 1) = 3\mathbf{i} + 9\mathbf{j}$,
$D_{\mathbf{u}}f = \nabla f \cdot \mathbf{u} = 12/\sqrt{2} = 6\sqrt{2}$

3. $\nabla f(x,y) = \left[2x/\left(1 + x^2 + y\right)\right]\mathbf{i} + \left[1/\left(1 + x^2 + y\right)\right]\mathbf{j}$, $\nabla f(0,0) = \mathbf{j}$, $D_{\mathbf{u}}f = -3/\sqrt{10}$

5. $\nabla f(x,y,z) = 20x^4y^2z^3\mathbf{i} + 8x^5yz^3\mathbf{j} + 12x^5y^2z^2\mathbf{k}$, $\nabla f(2,-1,1) = 320\mathbf{i} - 256\mathbf{j} + 384\mathbf{k}$, $D_{\mathbf{u}}f = -320$

7. $\nabla f(x,y,z) = \dfrac{2x}{x^2 + 2y^2 + 3z^2}\mathbf{i} + \dfrac{4y}{x^2 + 2y^2 + 3z^2}\mathbf{j} + \dfrac{6z}{x^2 + 2y^2 + 3z^2}\mathbf{k}$,

$\nabla f(-1,2,4) = (-2/57)\mathbf{i} + (8/57)\mathbf{j} + (24/57)\mathbf{k}$, $D_{\mathbf{u}}f = -314/741$

9. $\nabla f(x,y) = 12x^2y^2\mathbf{i} + 8x^3y\mathbf{j}$, $\nabla f(2,1) = 48\mathbf{i} + 64\mathbf{j}$, $\mathbf{u} = (4/5)\mathbf{i} - (3/5)\mathbf{j}$, $D_{\mathbf{u}}f = \nabla f \cdot \mathbf{u} = 0$

11. $\nabla f(x,y) = \left(y^2/x\right)\mathbf{i} + 2y\ln x\mathbf{j}$, $\nabla f(1,4) = 16\mathbf{i}$, $\mathbf{u} = (-\mathbf{i} + \mathbf{j})/\sqrt{2}$, $D_{\mathbf{u}}f = -8\sqrt{2}$

13. $\nabla f(x,y) = -\left[y/\left(x^2 + y^2\right)\right]\mathbf{i} + \left[x/\left(x^2 + y^2\right)\right]\mathbf{j}$,

$\nabla f(-2,2) = -(\mathbf{i} + \mathbf{j})/4$, $\mathbf{u} = -(\mathbf{i} + \mathbf{j})/\sqrt{2}$, $D_{\mathbf{u}}f = \sqrt{2}/4$

15. $\nabla f(x,y,z) = \left(3x^2z - 2xy\right)\mathbf{i} - x^2\mathbf{j} + \left(x^3 + 2z\right)\mathbf{k}$, $\nabla f(2,-1,1) = 16\mathbf{i} - 4\mathbf{j} + 10\mathbf{k}$,

$\mathbf{u} = (3\mathbf{i} - \mathbf{j} + 2\mathbf{k})/\sqrt{14}$, $D_{\mathbf{u}}f = 72/\sqrt{14}$

17. $\nabla f(x,y,z) = -\dfrac{1}{z+y}\mathbf{i} - \dfrac{z-x}{(z+y)^2}\mathbf{j} + \dfrac{y+x}{(z+y)^2}\mathbf{k}$, $\nabla f(1,0,-3) = (1/3)\mathbf{i} + (4/9)\mathbf{j} + (1/9)\mathbf{k}$,

$\mathbf{u} = (-6\mathbf{i} + 3\mathbf{j} - 2\mathbf{k})/7$, $D_{\mathbf{u}}f = -8/63$

19. $\nabla f(x,y) = (y/2)(xy)^{-1/2}\mathbf{i} + (x/2)(xy)^{-1/2}\mathbf{j}$, $\nabla f(1,4) = \mathbf{i} + (1/4)\mathbf{j}$,

$\mathbf{u} = \cos\theta\mathbf{i} + \sin\theta\mathbf{j} = (1/2)\mathbf{i} + (\sqrt{3}/2)\mathbf{j}$, $D_{\mathbf{u}}f = 1/2 + \sqrt{3}/8$

21. $\nabla f(x,y) = 2\sec^2(2x+y)\mathbf{i} + \sec^2(2x+y)\mathbf{j}$, $\nabla f(\pi/6,\pi/3) = 8\mathbf{i} + 4\mathbf{j}$, $\mathbf{u} = (\mathbf{i} - \mathbf{j})/\sqrt{2}$, $D_{\mathbf{u}}f = 2\sqrt{2}$

23. $\nabla f(x,y) = y(x+y)^{-2}\mathbf{i} - x(x+y)^{-2}\mathbf{j}$, $\nabla f(1,0) = -\mathbf{j}$, $\overrightarrow{PQ} = -2\mathbf{i} - \mathbf{j}$, $\mathbf{u} = (-2\mathbf{i} - \mathbf{j})/\sqrt{5}$,

$D_{\mathbf{u}}f = 1/\sqrt{5}$

25. $\nabla f(x,y) = \dfrac{ye^y}{2\sqrt{xy}}\mathbf{i} + \left(\sqrt{xy}e^y + \dfrac{xe^y}{2\sqrt{xy}}\right)\mathbf{j}$, $\nabla f(1,1) = (e/2)(\mathbf{i} + 3\mathbf{j})$, $\mathbf{u} = -\mathbf{j}$, $D_{\mathbf{u}}f = -3e/2$

27. $\nabla f(2,1,-1) = -\mathbf{i} + \mathbf{j} - \mathbf{k}$. $\overrightarrow{PQ} = -3\mathbf{i} + \mathbf{j} + \mathbf{k}$, $\mathbf{u} = (-3\mathbf{i} + \mathbf{j} + \mathbf{k})/\sqrt{11}$, $D_{\mathbf{u}}f = 3/\sqrt{11}$

29. Solve the system $(3/5)f_x(1,2) - (4/5)f_y(1,2) = -5$, $(4/5)f_x(1,2) + (3/5)f_y(1,2) = 10$ for

   (a) $f_x(1,2) = 5$ **(b)** $f_y(1,2) = 10$

   (c) $\nabla f(1,2) = 5\mathbf{i} + 10\mathbf{j}$, $\mathbf{u} = (-\mathbf{i} - 2\mathbf{j})/\sqrt{5}$, $D_{\mathbf{u}}f = -5\sqrt{5}$.

31. f increases the most in the direction of III.

33. $\nabla z = 4\mathbf{i} - 8\mathbf{j}$

35. $\nabla w = \dfrac{x}{x^2 + y^2 + z^2}\mathbf{i} + \dfrac{y}{x^2 + y^2 + z^2}\mathbf{j} + \dfrac{z}{x^2 + y^2 + z^2}\mathbf{k}$

37. $\nabla f(x,y) = 3(2x + y)\left(x^2 + xy\right)^2\mathbf{i} + 3x\left(x^2 + xy\right)^2\mathbf{j}$, $\nabla f(-1,-1) = -36\mathbf{i} - 12\mathbf{j}$

39. $\nabla f(x,y,z) = [y/(x+y+z)]\mathbf{i} + [y/(x+y+z) + \ln(x+y+z)]\mathbf{j} + [y/(x+y+z)]\mathbf{k}$,

$\nabla f(-3,4,0) = 4\mathbf{i} + 4\mathbf{j} + 4\mathbf{k}$

41. $f(1,2) = 3$,
level curve $4x - 2y + 3 = 3$,
$2x - y = 0$;
$\nabla f(x,y) = 4\mathbf{i} - 2\mathbf{j}$
$\nabla f(1,2) = 4\mathbf{i} - 2\mathbf{j}$

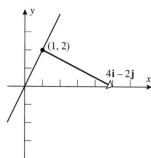

43. $f(-2,0) = 4$,
level curve $x^2 + 4y^2 = 4$,
$x^2/4 + y^2 = 1$.
$\nabla f(x,y) = 2x\mathbf{i} + 8y\mathbf{j}$
$\nabla f(-2,0) = -4\mathbf{i}$

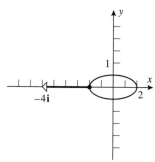

45. $\nabla f(x,y) = 8xy\mathbf{i} + 4x^2\mathbf{j}$, $\nabla f(1,-2) = -16\mathbf{i} + 4\mathbf{j}$ is normal to the level curve through P so
$\mathbf{u} = \pm(-4\mathbf{i} + \mathbf{j})/\sqrt{17}$.

47. $\nabla f(x,y) = 12x^2y^2\mathbf{i} + 8x^3y\mathbf{j}$, $\nabla f(-1,1) = 12\mathbf{i} - 8\mathbf{j}$, $\mathbf{u} = (3\mathbf{i} - 2\mathbf{j})/\sqrt{13}$, $\|\nabla f(-1,1)\| = 4\sqrt{13}$

49. $\nabla f(x,y) = x\left(x^2 + y^2\right)^{-1/2}\mathbf{i} + y\left(x^2 + y^2\right)^{-1/2}\mathbf{j}$,
$\nabla f(4,-3) = (4\mathbf{i} - 3\mathbf{j})/5$, $\mathbf{u} = (4\mathbf{i} - 3\mathbf{j})/5$, $\|\nabla f(4,-3)\| = 1$

51. $\nabla f(1,1,-1) = 3\mathbf{i} - 3\mathbf{j}$, $\mathbf{u} = (\mathbf{i} - \mathbf{j})/\sqrt{2}$, $\|\nabla f(1,1,-1)\| = 3\sqrt{2}$

53. $\nabla f(1,2,-2) = (-\mathbf{i} + \mathbf{j})/2$, $\mathbf{u} = (-\mathbf{i} + \mathbf{j})/\sqrt{2}$, $\|\nabla f(1,2,-2)\| = 1/\sqrt{2}$

55. $\nabla f(x,y) = -2x\mathbf{i} - 2y\mathbf{j}$, $\nabla f(-1,-3) = 2\mathbf{i} + 6\mathbf{j}$, $\mathbf{u} = -(\mathbf{i} + 3\mathbf{j})/\sqrt{10}$, $-\|\nabla f(-1,-3)\| = -2\sqrt{10}$

57. $\nabla f(x,y) = -3\sin(3x - y)\mathbf{i} + \sin(3x - y)\mathbf{j}$,
$\nabla f(\pi/6, \pi/4) = (-3\mathbf{i} + \mathbf{j})/\sqrt{2}$, $\mathbf{u} = (3\mathbf{i} - \mathbf{j})/\sqrt{10}$, $-\|\nabla f(\pi/6, \pi/4)\| = -\sqrt{5}$

59. $\nabla f(5,7,6) = -\mathbf{i} + 11\mathbf{j} - 12\mathbf{k}$, $\mathbf{u} = (\mathbf{i} - 11\mathbf{j} + 12\mathbf{k})/\sqrt{266}$, $-\|\nabla f(5,7,6)\| = -\sqrt{266}$

61. $\nabla f(4,-5) = 2\mathbf{i} - \mathbf{j}$, $\mathbf{u} = (5\mathbf{i} + 2\mathbf{j})/\sqrt{29}$, $D_{\mathbf{u}}f = 8/\sqrt{29}$

63. **(a)** At $(1,2)$ the steepest ascent seems to be in the direction $\mathbf{i} + \mathbf{j}$ and the slope in that direction seems to be $0.5/(\sqrt{2}/2) = 1/\sqrt{2}$, so $\nabla f \approx \frac{1}{2}\mathbf{i} + \frac{1}{2}\mathbf{j}$, which has the required direction and magnitude.

(b) The direction of $-\nabla f(4,4)$ appears to be $-\mathbf{i} - \mathbf{j}$ and its magnitude appears to be $1/0.8 = 5/4$.

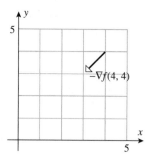

65. $\nabla z = 6x\mathbf{i} - 2y\mathbf{j}$, $\|\nabla z\| = \sqrt{36x^2 + 4y^2} = 6$ if $36x^2 + 4y^2 = 36$; all points on the ellipse $9x^2 + y^2 = 9$.

67. $\mathbf{r} = t\mathbf{i} - t^2\mathbf{j}$, $d\mathbf{r}/dt = \mathbf{i} - 2t\mathbf{j} = \mathbf{i} - 4\mathbf{j}$ at the point $(2, -4)$, $\mathbf{u} = (\mathbf{i} - 4\mathbf{j})/\sqrt{17}$;
$\nabla z = 2x\mathbf{i} + 2y\mathbf{j} = 4\mathbf{i} - 8\mathbf{j}$ at $(2, -4)$, hence $dz/ds = D_{\mathbf{u}}z = \nabla z \cdot \mathbf{u} = 36/\sqrt{17}$.

69. **(a)** $\nabla V(x, y) = -2e^{-2x}\cos 2y\mathbf{i} - 2e^{-2x}\sin 2y\mathbf{j}$, $\mathbf{E} = -\nabla V(\pi/4, 0) = 2e^{-\pi/2}\mathbf{i}$

   (b) $V(x, y)$ decreases most rapidly in the direction of $-\nabla V(x, y)$ which is $\mathbf{E}$.

71. Let $\mathbf{u}$ be the unit vector in the direction of $\mathbf{a}$, then
$D_{\mathbf{u}}f(3, -2, 1) = \nabla f(3, -2, 1) \cdot \mathbf{u} = \|\nabla f(3, -2, 1)\|\cos\theta = 5\cos\theta = -5$, $\cos\theta = -1$, $\theta = \pi$ so
$\nabla f(3, -2, 1)$ is oppositely directed to $\mathbf{u}$; $\nabla f(3, -2, 1) = -5\mathbf{u} = -10/3\mathbf{i} + 5/3\mathbf{j} + 10/3\mathbf{k}$.

73. **(a)** $\nabla r = \dfrac{x}{\sqrt{x^2 + y^2}}\mathbf{i} + \dfrac{y}{\sqrt{x^2 + y^2}}\mathbf{j} = \mathbf{r}/r$

   (b) $\nabla f(r) = \dfrac{\partial f(r)}{\partial x}\mathbf{i} + \dfrac{\partial f(r)}{\partial y}\mathbf{j} = f'(r)\dfrac{\partial r}{\partial x}\mathbf{i} + f'(r)\dfrac{\partial r}{\partial y}\mathbf{j} = f'(r)\nabla r$

75. $\mathbf{u}_r = \cos\theta\mathbf{i} + \sin\theta\mathbf{j}$, $\mathbf{u}_\theta = -\sin\theta\mathbf{i} + \cos\theta\mathbf{j}$,

$$\nabla z = \frac{\partial z}{\partial x}\mathbf{i} + \frac{\partial z}{\partial y}\mathbf{j} = \left(\frac{\partial z}{\partial r}\cos\theta - \frac{1}{r}\frac{\partial z}{\partial\theta}\sin\theta\right)\mathbf{i} + \left(\frac{\partial z}{\partial r}\sin\theta + \frac{1}{r}\frac{\partial z}{\partial\theta}\cos\theta\right)\mathbf{j}$$

$$= \frac{\partial z}{\partial r}(\cos\theta\mathbf{i} + \sin\theta\mathbf{j}) + \frac{1}{r}\frac{\partial z}{\partial\theta}(-\sin\theta\mathbf{i} + \cos\theta\mathbf{j}) = \frac{\partial z}{\partial r}\mathbf{u}_r + \frac{1}{r}\frac{\partial z}{\partial\theta}\mathbf{u}_\theta$$

77. $\mathbf{r}'(t) = \mathbf{v}(t) = k(x, y)\nabla T = -8k(x, y)x\mathbf{i} - 2k(x, y)y\mathbf{j}$; $\dfrac{dx}{dt} = -8kx$, $\dfrac{dy}{dt} = -2ky$. Divide and solve
to get $y^4 = 256x$; one parametrization is $x(t) = e^{-8t}$, $y(t) = 4e^{-2t}$.

79.

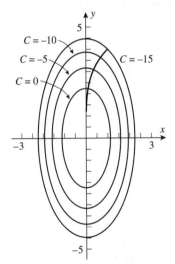

81. **(a)**

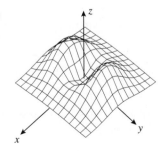

   (c) $\nabla f = [2x - 2x(x^2 + 3y^2)]e^{-(x^2+y^2)}\mathbf{i}$
   $\qquad\qquad + [6y - 2y(x^2 + 3y^2)]e^{-(x^2+y^2)}\mathbf{j}$

   (d) $\nabla f = \mathbf{0}$ if $x = y = 0$ or $x = 0, y = \pm 1$ or $x = \pm 1, y = 0$.

83. $\nabla f(x,y) = f_x(x,y)\mathbf{i} + f_y(x,y)\mathbf{j}$, if $\nabla f(x,y) = 0$ throughout the region then $f_x(x,y) = f_y(x,y) = 0$ throughout the region, the result follows from Exercise 71, Section 14.5.

85. $\nabla f(u,v,w) = \dfrac{\partial f}{\partial x}\mathbf{i} + \dfrac{\partial f}{\partial y}\mathbf{j} + \dfrac{\partial f}{\partial z}\mathbf{k}$

$$= \left(\frac{\partial f}{\partial u}\frac{\partial u}{\partial x} + \frac{\partial f}{\partial v}\frac{\partial v}{\partial x} + \frac{\partial f}{\partial w}\frac{\partial w}{\partial x}\right)\mathbf{i} + \left(\frac{\partial f}{\partial u}\frac{\partial u}{\partial y} + \frac{\partial f}{\partial v}\frac{\partial v}{\partial y} + \frac{\partial f}{\partial w}\frac{\partial w}{\partial y}\right)\mathbf{j}$$

$$+ \left(\frac{\partial f}{\partial u}\frac{\partial u}{\partial z} + \frac{\partial f}{\partial v}\frac{\partial v}{\partial z} + \frac{\partial f}{\partial w}\frac{\partial w}{\partial z}\right)\mathbf{k} = \frac{\partial f}{\partial u}\nabla u + \frac{\partial f}{\partial v}\nabla v + \frac{\partial f}{\partial w}\nabla w$$

87. (a) $\dfrac{d}{ds}f(x_0+su_1, y_0+su_2)$ at $s = 0$ is by definition equal to $\displaystyle\lim_{s\to 0}\dfrac{f(x_0 + su_1, y_0 + su_2) - f(x_0, y_0)}{s}$, and from Exercise 86(a) this value is equal to $f_x(x_0, y_0)u_1 + f_y(x_0, y_0)u_2$.

(b) For any number $\epsilon > 0$ a number $\delta > 0$ exists such that whenever $0 < |s| < \delta$ then
$$\left|\frac{f(x_0 + su_1, y_0 + su_2) - f(x_0, y_0) - f_x(x_0, y_0)su_1 - f_y(x_0, y_0)su_2}{s}\right| < \epsilon.$$

(c) For any number $\epsilon > 0$ there exists a number $\delta > 0$ such that $\dfrac{|E(x,y)|}{\sqrt{(x - x_0)^2 + (y - y_0)^2}} < \epsilon$ whenever $0 < \sqrt{(x - x_0)^2 + (y - y_0)^2} < \delta$.

(d) For any number $\epsilon > 0$ there exists a number $\delta > 0$ such that
$$\left|\frac{f(x_0 + su_1, y_0 + su_2) - f(x_0, y_0) - f_x(x_0, y_0)su_1 - f_y(x_0, y_0)su_2}{s}\right| < \epsilon \text{ when } 0 < |s| < \delta.$$

(e) Since f is differentiable at (x_0, y_0), by Part (c) the Equation (5) of Definition 14.2.1 holds. By Part (d), for any $\epsilon > 0$ there exists $\delta > 0$ such that
$$\left|\frac{f(x_0 + su_1, y_0 + su_2) - f(x_0, y_0) - f_x(x_0, y_0)su_1 - f_y(x_0, y_0)su_2}{s}\right| < \epsilon \text{ when } 0 < |s| < \delta.$$
By Part (a) it follows that the limit in Part (a) holds, and thus that
$$\frac{d}{ds}f(x_0 + su_1, y_0 + su_2)\Big]_{s=0} = f_x(x_0, y_0)u_1 + f_y(x_0, y_0)u_2,$$
which proves Equation (4) of Theorem 14.6.3.

EXERCISE SET 14.7

1. At P, $\partial z/\partial x = 48$ and $\partial z/\partial y = -14$, tangent plane $48x - 14y - z = 64$, normal line $x = 1 + 48t$, $y = -2 - 14t$, $z = 12 - t$.

3. At P, $\partial z/\partial x = 1$ and $\partial z/\partial y = -1$, tangent plane $x - y - z = 0$, normal line $x = 1 + t$, $y = -t$, $z = 1 - t$.

5. At P, $\partial z/\partial x = 0$ and $\partial z/\partial y = 3$, tangent plane $3y - z = -1$, normal line $x = \pi/6$, $y = 3t$, $z = 1 - t$.

7. By implicit differentiation $\partial z/\partial x = -x/z$, $\partial z/\partial y = -y/z$ so at P, $\partial z/\partial x = 3/4$ and $\partial z/\partial y = 0$, tangent plane $3x - 4z = -25$, normal line $x = -3 + 3t/4$, $y = 0$, $z = 4 - t$.

9. The tangent plane is horizontal if the normal $\partial z/\partial x\mathbf{i} + \partial z/\partial y\mathbf{j} - \mathbf{k}$ is parallel to $\mathbf{k}$ which occurs when $\partial z/\partial x = \partial z/\partial y = 0$.

(a) $\partial z/\partial x = 3x^2y^2$, $\partial z/\partial y = 2x^3y$; $3x^2y^2 = 0$ and $2x^3y = 0$ for all (x,y) on the x-axis or y-axis, and $z = 0$ for these points, the tangent plane is horizontal at all points on the x-axis or y-axis.

(b) $\partial z/\partial x = 2x - y - 2$, $\partial z/\partial y = -x + 2y + 4$; solve the system $2x - y - 2 = 0$, $-x + 2y + 4 = 0$, to get $x = 0$, $y = -2$. $z = -4$ at $(0, -2)$, the tangent plane is horizontal at $(0, -2, -4)$.

11. $\partial z/\partial x = -6x$, $\partial z/\partial y = -4y$ so $-6x_0\mathbf{i} - 4y_0\mathbf{j} - \mathbf{k}$ is normal to the surface at a point (x_0, y_0, z_0) on the surface. This normal must be parallel to the given line and hence to the vector $-3\mathbf{i} + 8\mathbf{j} - \mathbf{k}$ which is parallel to the line so $-6x_0 = -3$, $x_0 = 1/2$ and $-4y_0 = 8$, $y_0 = -2$. $z = -3/4$ at $(1/2, -2)$. The point on the surface is $(1/2, -2, -3/4)$.

13. (a) $2t + 7 = (-1 + t)^2 + (2 + t)^2$, $t^2 = 1$, $t = \pm 1$ so the points of intersection are $(-2, 1, 5)$ and $(0, 3, 9)$.

(b) $\partial z/\partial x = 2x$, $\partial z/\partial y = 2y$ so at $(-2, 1, 5)$ the vector $\mathbf{n} = -4\mathbf{i} + 2\mathbf{j} - \mathbf{k}$ is normal to the surface. $\mathbf{v} = \mathbf{i} + \mathbf{j} + 2\mathbf{k}$ is parallel to the line; $\mathbf{n} \cdot \mathbf{v} = -4$ so the cosine of the acute angle is $[\mathbf{n} \cdot (-\mathbf{v})]/(\|\mathbf{n}\| \| - \mathbf{v}\|) = 4/\left(\sqrt{21}\sqrt{6}\right) = 4/\left(3\sqrt{14}\right)$. Similarly, at $(0,3,9)$ the vector $\mathbf{n} = 6\mathbf{j} - \mathbf{k}$ is normal to the surface, $\mathbf{n} \cdot \mathbf{v} = 4$ so the cosine of the acute angle is $4/\left(\sqrt{37}\sqrt{6}\right) = 4/\sqrt{222}$.

15. (a) $f(x, y, z) = x^2 + y^2 + 4z^2$, $\nabla f = 2x\mathbf{i} + 2y\mathbf{j} + 8z\mathbf{k}$, $\nabla f(2, 2, 1) = 4\mathbf{i} + 4\mathbf{j} + 8\mathbf{k}$, $\mathbf{n} = \mathbf{i} + \mathbf{j} + 2\mathbf{k}$, $x + y + 2z = 6$

(b) $\mathbf{r}(t) = 2\mathbf{i} + 2\mathbf{j} + \mathbf{k} + t(\mathbf{i} + \mathbf{j} + 2\mathbf{k})$, $x(t) = 2 + t$, $y(t) = 2 + t$, $z(t) = 1 + 2t$

(c) $\cos\theta = \dfrac{\mathbf{n} \cdot \mathbf{k}}{\|\mathbf{n}\|} = \dfrac{\sqrt{2}}{\sqrt{3}}$, $\theta \approx 35.26°$

17. Set $f(x, y) = z + x - z^4(y - 1)$, then $f(x, y, z) = 0$, $\mathbf{n} = \pm\nabla f(3, 5, 1) = \pm(\mathbf{i} - \mathbf{j} - 19\mathbf{k})$,

unit vectors $\pm\dfrac{1}{\sqrt{363}}(\mathbf{i} - \mathbf{j} - 19\mathbf{k})$

19. $f(x, y, z) = x^2 + y^2 + z^2$, if (x_0, y_0, z_0) is on the sphere then $\nabla f(x_0, y_0, z_0) = 2(x_0\mathbf{i} + y_0\mathbf{j} + z_0\mathbf{k})$ is normal to the sphere at (x_0, y_0, z_0), the normal line is $x = x_0 + x_0t$, $y = y_0 + y_0t$, $z = z_0 + z_0t$ which passes through the origin when $t = -1$.

21. $f(x, y, z) = x^2 + y^2 - z^2$, if (x_0, y_0, z_0) is on the surface then $\nabla f(x_0, y_0, z_0) = 2(x_0\mathbf{i} + y_0\mathbf{j} - z_0\mathbf{k})$ is normal there and hence so is $\mathbf{n}_1 = x_0\mathbf{i} + y_0\mathbf{j} - z_0\mathbf{k}$; $\mathbf{n}_1$ must be parallel to $\overrightarrow{PQ} = 3\mathbf{i} + 2\mathbf{j} - 2\mathbf{k}$ so $\mathbf{n}_1 = c\overrightarrow{PQ}$ for some constant c. Equate components to get $x_0 = 3c$, $y_0 = 2c$ and $z_0 = 2c$ which when substituted into the equation of the surface yields $9c^2 + 4c^2 - 4c^2 = 1$, $c^2 = 1/9$, $c = \pm 1/3$ so the points are $(1, 2/3, 2/3)$ and $(-1, -2/3, -2/3)$.

23. $\mathbf{n}_1 = 2\mathbf{i} - 2\mathbf{j} - \mathbf{k}$, $\mathbf{n}_2 = 2\mathbf{i} - 8\mathbf{j} + 4\mathbf{k}$, $\mathbf{n}_1 \times \mathbf{n}_2 = -16\mathbf{i} - 10\mathbf{j} - 12\mathbf{k}$ is tangent to the line, so $x(t) = 1 + 8t$, $y(t) = -1 + 5t$, $z(t) = 2 + 6t$

25. $f(x, y, z) = x^2 + z^2 - 25$, $g(x, y, z) = y^2 + z^2 - 25$, $\mathbf{n}_1 = \nabla f(3, -3, 4) = 6\mathbf{i} + 8\mathbf{k}$, $\mathbf{n}_2 = \nabla g(3, -3, 4) = -6\mathbf{j} + 8\mathbf{k}$, $\mathbf{n}_1 \times \mathbf{n}_2 = 48\mathbf{i} - 48\mathbf{j} - 36\mathbf{k}$ is tangent to the line, $x(t) = 3 + 4t$, $y(t) = -3 - 4t$, $z(t) = 4 - 3t$

27. Use implicit differentiation to get $\partial z/\partial x = -c^2x/\left(a^2z\right)$, $\partial z/\partial y = -c^2y/\left(b^2z\right)$. At (x_0, y_0, z_0), $z_0 \neq 0$, a normal to the surface is $-\left[c^2x_0/\left(a^2z_0\right)\right]\mathbf{i} - \left[c^2y_0/\left(b^2z_0\right)\right]\mathbf{j} - \mathbf{k}$ so the tangent plane is

$$-\frac{c^2x_0}{a^2z_0}x - \frac{c^2y_0}{b^2z_0}y - z = -\frac{c^2x_0^2}{a^2z_0} - \frac{c^2y_0^2}{b^2z_0} - z_0, \quad \frac{x_0x}{a^2} + \frac{y_0y}{b^2} + \frac{z_0z}{c^2} = \frac{x_0^2}{a^2} + \frac{y_0^2}{b^2} + \frac{z_0^2}{c^2} = 1$$

29. $n_1 = f_x(x_0, y_0) i + f_y(x_0, y_0) j - k$ and $n_2 = g_x(x_0, y_0) i + g_y(x_0, y_0) j - k$ are normal, respectively, to $z = f(x, y)$ and $z = g(x, y)$ at P; n_1 and n_2 are perpendicular if and only if $n_1 \cdot n_2 = 0$, $f_x(x_0, y_0) g_x(x_0, y_0) + f_y(x_0, y_0) g_y(x_0, y_0) + 1 = 0$, $f_x(x_0, y_0) g_x(x_0, y_0) + f_y(x_0, y_0) g_y(x_0, y_0) = -1$.

31. $\nabla f = f_x i + f_y j + f_z k$ and $\nabla g = g_x i + g_y j + g_z k$ evaluated at (x_0, y_0, z_0) are normal, respectively, to the surfaces $f(x, y, z) = 0$ and $g(x, y, z) = 0$ at (x_0, y_0, z_0). The surfaces are orthogonal at (x_0, y_0, z_0) if and only if $\nabla f \cdot \nabla g = 0$ so $f_x g_x + f_y g_y + f_z g_z = 0$.

33. $z = \dfrac{k}{xy}$; at a point $\left(a, b, \dfrac{k}{ab}\right)$ on the surface, $\left\langle -\dfrac{k}{a^2 b}, -\dfrac{k}{ab^2}, -1 \right\rangle$ and hence $\langle bk, ak, a^2 b^2 \rangle$ is normal to the surface so the tangent plane is $bkx + aky + a^2 b^2 z = 3abk$. The plane cuts the x, y, and z-axes at the points $3a$, $3b$, and $\dfrac{3k}{ab}$, respectively, so the volume of the tetrahedron that is formed is $V = \dfrac{1}{3}\left(\dfrac{3k}{ab}\right)\left[\dfrac{1}{2}(3a)(3b)\right] = \dfrac{9}{2}k$, which does not depend on a and b.

EXERCISE SET 14.8

1. **(a)** minimum at $(2, -1)$, no maxima   **(b)** maximum at $(0, 0)$, no minima
   (c) no maxima or minima

3. $f(x, y) = (x - 3)^2 + (y + 2)^2$, minimum at $(3, -2)$, no maxima

5. $f_x = 6x + 2y = 0$, $f_y = 2x + 2y = 0$; critical point $(0,0)$; $D = 8 > 0$ and $f_{xx} = 6 > 0$ at $(0,0)$, relative minimum.

7. $f_x = 2x - 2xy = 0$, $f_y = 4y - x^2 = 0$; critical points $(0,0)$ and $(\pm 2, 1)$; $D = 8 > 0$ and $f_{xx} = 2 > 0$ at $(0,0)$, relative minimum; $D = -16 < 0$ at $(\pm 2, 1)$, saddle points.

9. $f_x = y + 2 = 0$, $f_y = 2y + x + 3 = 0$; critical point $(1, -2)$; $D = -1 < 0$ at $(1, -2)$, saddle point.

11. $f_x = 2x + y - 3 = 0$, $f_y = x + 2y = 0$; critical point $(2, -1)$; $D = 3 > 0$ and $f_{xx} = 2 > 0$ at $(2, -1)$, relative minimum.

13. $f_x = 2x - 2/(x^2 y) = 0$, $f_y = 2y - 2/(xy^2) = 0$; critical points $(-1, -1)$ and $(1, 1)$; $D = 32 > 0$ and $f_{xx} = 6 > 0$ at $(-1, -1)$ and $(1, 1)$, relative minima.

15. $f_x = 2x = 0$, $f_y = 1 - e^y = 0$; critical point $(0, 0)$; $D = -2 < 0$ at $(0, 0)$, saddle point.

17. $f_x = e^x \sin y = 0$, $f_y = e^x \cos y = 0$, $\sin y = \cos y = 0$ is impossible, no critical points.

19. $f_x = -2(x + 1)e^{-(x^2 + y^2 + 2x)} = 0$, $f_y = -2y e^{-(x^2 + y^2 + 2x)} = 0$; critical point $(-1, 0)$; $D = 4e^2 > 0$ and $f_{xx} = -2e < 0$ at $(-1, 0)$, relative maximum.

21.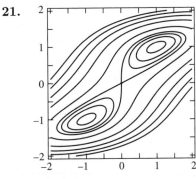

$\nabla f = (4x - 4y)\mathbf{i} - (4x - 4y^3)\mathbf{j} = \mathbf{0}$ when $x = y, x = y^3$, so $x = y = 0$ or $x = y = \pm 1$. At $(0,0), D = -16$, a saddle point; at $(1,1)$ and $(-1,-1), D = 32 > 0, f_{xx} = 4$, a relative minimum.

23. (a) critical point $(0,0)$; $D = 0$

    (b) $f(0,0) = 0$, $x^4 + y^4 \geq 0$ so $f(x,y) \geq f(0,0)$, relative minimum.

25. (a) $f_x = 3e^y - 3x^2 = 3\left(e^y - x^2\right) = 0$, $f_y = 3xe^y - 3e^{3y} = 3e^y\left(x - e^{2y}\right) = 0$, $e^y = x^2$ and $e^{2y} = x$, $x^4 = x$, $x\left(x^3 - 1\right) = 0$ so $x = 0, 1$; critical point $(1,0)$; $D = 27 > 0$ and $f_{xx} = -6 < 0$ at $(1,0)$, relative maximum.

    (b) $\lim\limits_{x \to -\infty} f(x,0) = \lim\limits_{x \to -\infty}\left(3x - x^3 - 1\right) = +\infty$ so no absolute maximum.

27. $f_x = y - 1 = 0$, $f_y = x - 3 = 0$; critical point $(3,1)$.

        Along $y = 0$: $u(x) = -x$; no critical points,

        along $x = 0$: $v(y) = -3y$; no critical points,

    along $y = -\dfrac{4}{5}x + 4$: $w(x) = -\dfrac{4}{5}x^2 + \dfrac{27}{5}x - 12$; critical point $(27/8, 13/10)$.

(x,y)	$(3,1)$	$(0,0)$	$(5,0)$	$(0,4)$	$(27/8, 13/10)$
$f(x,y)$	-3	0	-5	-12	$-231/80$

Absolute maximum value is 0, absolute minimum value is -12.

29. $f_x = 2x - 2 = 0$, $f_y = -6y + 6 = 0$; critical point $(1,1)$.

    Along $y = 0$: $u_1(x) = x^2 - 2x$; critical point $(1,0)$,

    along $y = 2$: $u_2(x) = x^2 - 2x$; critical point $(1,2)$,

    along $x = 0$: $v_1(y) = -3y^2 + 6y$; critical point $(0,1)$,

    along $x = 2$: $v_2(y) = -3y^2 + 6y$; critical point $(2,1)$

(x,y)	$(1,1)$	$(1,0)$	$(1,2)$	$(0,1)$	$(2,1)$	$(0,0)$	$(0,2)$	$(2,0)$	$(2,2)$
$f(x,y)$	2	-1	-1	3	3	0	0	0	0

Absolute maximum value is 3, absolute minimum value is -1.

31. $f_x = 2x - 1 = 0$, $f_y = 4y = 0$; critical point $(1/2, 0)$.

    Along $x^2 + y^2 = 4 : y^2 = 4 - x^2$, $u(x) = 8 - x - x^2$ for $-2 \leq x \leq 2$; critical points $(-1/2, \pm\sqrt{15}/2)$.

(x,y)	$(1/2,0)$	$(-1/2, \sqrt{15}/2)$	$(-1/2, -\sqrt{15}/2)$	$(-2,0)$	$(2,0)$
$f(x,y)$	$-1/4$	$33/4$	$33/4$	6	2

Absolute maximum value is $33/4$, absolute minimum value is $-1/4$.

33. Maximize $P = xyz$ subject to $x + y + z = 48$, $x > 0$, $y > 0$, $z > 0$. $z = 48 - x - y$ so
$P = xy(48 - x - y) = 48xy - x^2y - xy^2$, $P_x = 48y - 2xy - y^2 = 0$, $P_y = 48x - x^2 - 2xy = 0$. But
$x \neq 0$ and $y \neq 0$ so $48 - 2x - y = 0$ and $48 - x - 2y = 0$; critical point $(16,16)$. $P_{xx}P_{yy} - P_{xy}^2 > 0$
and $P_{xx} < 0$ at $(16, 16)$, relative maximum. $z = 16$ when $x = y = 16$, the product is maximum for
the numbers $16, 16, 16$.

35. Maximize $w = xy^2z^2$ subject to $x + y + z = 5$, $x > 0$, $y > 0$, $z > 0$. $x = 5 - y - z$ so
$w = (5 - y - z)y^2z^2 = 5y^2z^2 - y^3z^2 - y^2z^3$, $w_y = 10yz^2 - 3y^2z^2 - 2yz^3 = yz^2(10 - 3y - 2z) = 0$,
$w_z = 10y^2z - 2y^3z - 3y^2z^2 = y^2z(10 - 2y - 3z) = 0$, $10 - 3y - 2z = 0$ and $10 - 2y - 3z = 0$; critical
point when $y = z = 2$; $w_{yy}w_{zz} - w_{yz}^2 = 320 > 0$ and $w_{yy} = -24 < 0$ when $y = z = 2$, relative
maximum. $x = 1$ when $y = z = 2$, xy^2z^2 is maximum at $(1, 2, 2)$.

37. The diagonal of the box must equal the diameter of the sphere, thus we maximize $V = xyz$ or, for
convenience, $w = V^2 = x^2y^2z^2$ subject to $x^2 + y^2 + z^2 = 4a^2$, $x > 0$, $y > 0$, $z > 0$; $z^2 = 4a^2 - x^2 - y^2$
hence $w = 4a^2x^2y^2 - x^4y^2 - x^2y^4$, $w_x = 2xy^2(4a^2 - 2x^2 - y^2) = 0$, $w_y = 2x^2y\left(4a^2 - x^2 - 2y^2\right) = 0$,
$4a^2 - 2x^2 - y^2 = 0$ and $4a^2 - x^2 - 2y^2 = 0$; critical point $\left(2a/\sqrt{3}, 2a/\sqrt{3}\right)$;
$w_{xx}w_{yy} - w_{xy}^2 = \dfrac{4096}{27}a^8 > 0$ and $w_{xx} = -\dfrac{128}{9}a^4 < 0$ at $\left(2a/\sqrt{3}, 2a/\sqrt{3}\right)$, relative maximum.
$z = 2a/\sqrt{3}$ when $x = y = 2a/\sqrt{3}$, the dimensions of the box of maximum volume are
$2a/\sqrt{3}, 2a/\sqrt{3}, 2a/\sqrt{3}$.

39. Let x, y, and z be, respectively, the length, width, and height of the box. Minimize
$C = 10(2xy) + 5(2xz + 2yz) = 10(2xy + xz + yz)$ subject to $xyz = 16$. $z = 16/(xy)$
so $C = 20(xy + 8/y + 8/x)$, $C_x = 20(y - 8/x^2) = 0$, $C_y = 20(x - 8/y^2) = 0$;
critical point $(2,2)$; $C_{xx}C_{yy} - C_{xy}^2 = 1200 > 0$
and $C_{xx} = 40 > 0$ at $(2,2)$, relative minimum. $z = 4$ when $x = y = 2$. The cost of materials is
minimum if the length and width are 2 ft and the height is 4 ft.

41. (a) $x = 0 : f(0, y) = -3y^2$, minimum -3, maximum 0;

$x = 1, f(1, y) = 4 - 3y^2 + 2y, \dfrac{\partial f}{\partial y}(1, y) = -6y + 2 = 0$ at $y = 1/3$, minimum 3,

maximum $13/3$;

$y = 0, f(x, 0) = 4x^2$, minimum 0, maximum 4;

$y = 1, f(x, 1) = 4x^2 + 2x - 3, \dfrac{\partial f}{\partial x}(x, 1) = 8x + 2 \neq 0$ for $0 < x < 1$, minimum -3, maximum 3

(b) $f(x, x) = 3x^2$, minimum 0, maximum 3; $f(x, 1-x) = -x^2 + 8x - 3, \dfrac{d}{dx}f(x, 1-x) = -2x + 8 \neq 0$
for $0 < x < 1$, maximum 4, minimum -3

(c) $f_x(x, y) = 8x + 2y = 0, f_y(x, y) = -6y + 2x = 0$, solution is $(0, 0)$, which is not an interior
point of the square, so check the sides: minimum -3, maximum $13/3$.

43. Minimize $S = xy + 2xz + 2yz$ subject to $xyz = V$, $x > 0$, $y > 0$, $z > 0$ where x, y, and z are,
respectively, the length, width, and height of the box. $z = V/(xy)$ so $S = xy + 2V/y + 2V/x$,
$S_x = y - 2V/x^2 = 0$, $S_y = x - 2V/y^2 = 0$; critical point $\left(\sqrt[3]{2V}, \sqrt[3]{2V}\right)$; $S_{xx}S_{yy} - S_{xy}^2 = 3 > 0$ and
$S_{xx} = 2 > 0$ at this point so there is a relative minimum there. The length and width are each
$\sqrt[3]{2V}$, the height is $z = \sqrt[3]{2V}/2$.

45. (a) $\dfrac{\partial g}{\partial m} = \displaystyle\sum_{i=1}^{n} 2\left(mx_i + b - y_i\right)x_i = 2\left(m\sum_{i=1}^{n} x_i^2 + b\sum_{i=1}^{n} x_i - \sum_{i=1}^{n} x_iy_i\right) = 0$ if

$\left(\displaystyle\sum_{i=1}^{n} x_i^2\right)m + \left(\sum_{i=1}^{n} x_i\right)b = \sum_{i=1}^{n} x_iy_i,$

$$\frac{\partial g}{\partial b} = \sum_{i=1}^{n} 2\left(mx_i + b - y_i\right) = 2\left(m\sum_{i=1}^{n} x_i + bn - \sum_{i=1}^{n} y_i\right) = 0 \text{ if } \left(\sum_{i=1}^{n} x_i\right)m + nb = \sum_{i=1}^{n} y_i$$

(b) $\displaystyle\sum_{i=1}^{n}(x_i - \bar{x})^2 = \sum_{i=1}^{n}\left(x_i^2 - 2\bar{x}x_i + \bar{x}^2\right) = \sum_{i=1}^{n} x_i^2 - 2\bar{x}\sum_{i=1}^{n} x_i + n\bar{x}^2$

$$= \sum_{i=1}^{n} x_i^2 - \frac{2}{n}\left(\sum_{i=1}^{n} x_i\right)^2 + \frac{1}{n}\left(\sum_{i=1}^{n} x_i\right)^2$$

$$= \sum_{i=1}^{n} x_i^2 - \frac{1}{n}\left(\sum_{i=1}^{n} x_i\right)^2 \geq 0 \text{ so } n\sum_{i=1}^{n} x_i^2 - \left(\sum_{i=1}^{n} x_i\right)^2 \geq 0$$

This is an equality if and only if $\displaystyle\sum_{i=1}^{n}(x_i - \bar{x})^2 = 0$, which means $x_i = \bar{x}$ for each i.

(c) The system of equations $Am + Bb = C, Dm + Eb = F$ in the unknowns m and b has a unique solution provided $AE \neq BD$, and if so the solution is $m = \dfrac{CE - BF}{AE - BD}, b = \dfrac{F - Dm}{E}$, which after the appropriate substitution yields the desired result.

47. $n = 3, \displaystyle\sum_{i=1}^{3} x_i = 3, \sum_{i=1}^{3} y_i = 7, \sum_{i=1}^{3} x_i y_i = 13, \sum_{i=1}^{3} x_i^2 = 11, y = \frac{3}{4}x + \frac{19}{12}$

49. $\displaystyle\sum_{i=1}^{4} x_i = 10, \sum_{i=1}^{4} y_i = 8.2, \sum_{i=1}^{4} x_i^2 = 30, \sum_{i=1}^{4} x_i y_i = 23, n = 4; \; m = 0.5, b = 0.8, y = 0.5x + 0.8.$

51. (a) $y = \dfrac{8843}{140} + \dfrac{57}{200}t \approx 63.1643 + 0.285t$

(b)

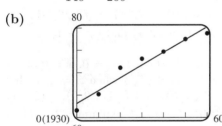

(c) $y = \dfrac{2909}{35} \approx 83.1143$

53. (a) $P = \dfrac{2798}{21} + \dfrac{171}{350}T \approx 133.2381 + 0.4886T$

(b)

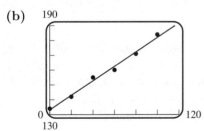

(c) $T \approx -\dfrac{139{,}900}{513} \approx -272.7096°\text{ C}$

55. $f(x_0, y_0) \geq f(x, y)$ for all (x, y) inside a circle centered at (x_0, y_0) by virtue of Definition 14.8.1. If r is the radius of the circle, then in particular $f(x_0, y_0) \geq f(x, y_0)$ for all x satisfying

$|x - x_0| < r$ so $f(x, y_0)$ has a relative maximum at x_0. The proof is similar for the function $f(x_0, y)$.

EXERCISE SET 14.9

1. **(a)** $xy = 4$ is tangent to the line, so the maximum value of f is 4.

   (b) $xy = 2$ intersects the curve and so gives a smaller value of f.

   (c) Maximize $f(x, y) = xy$ subject to the constraint $g(x, y) = x + y - 4 = 0, \nabla f = \lambda \nabla g$, $y\mathbf{i} + x\mathbf{j} = \lambda(\mathbf{i} + \mathbf{j})$, so solve the equations $y = \lambda, x = \lambda$ with solution $x = y = \lambda$, but $x + y = 4$, so $x = y = 2$, and the maximum value of f is $f = xy = 4$.

3. **(a)**

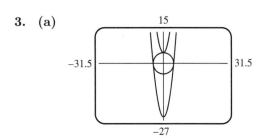

   (b) one extremum at $(0, 5)$ and one at approximately $(\pm 5, 0)$, so minimum value -5, maximum value ≈ 25

   (c) Find the minimum and maximum values of $f(x, y) = x^2 - y$ subject to the constraint $g(x, y) = x^2 + y^2 - 25 = 0, \nabla f = \lambda \nabla g, 2x\mathbf{i} - \mathbf{j} = 2\lambda x\mathbf{i} + 2\lambda y\mathbf{j}$, so solve $2x = 2\lambda x, -1 = 2\lambda y, x^2 + y^2 - 25 = 0$. If $x = 0$ then $y = \pm 5, f = \mp 5$, and if $x \neq 0$ then $\lambda = 1, y = -1/2, x^2 = 25 - 1/4 = 99/4, f = 99/4 + 1/2 = 101/4$, so the maximum value of f is $101/4$ at $(\pm 3\sqrt{11}/2, -1/2)$ and the minimum value of f is -5 at $(0, 5)$.

5. $y = 8x\lambda, x = 16y\lambda; y/(8x) = x/(16y), x^2 = 2y^2$ so $4\left(2y^2\right) + 8y^2 = 16, y^2 = 1, y = \pm 1$. Test $(\pm\sqrt{2}, -1)$ and $(\pm\sqrt{2}, 1)$. $f\left(-\sqrt{2}, -1\right) = f\left(\sqrt{2}, 1\right) = \sqrt{2}, f\left(-\sqrt{2}, 1\right) = f\left(\sqrt{2}, -1\right) = -\sqrt{2}$. Maximum $\sqrt{2}$ at $\left(-\sqrt{2}, -1\right)$ and $\left(\sqrt{2}, 1\right)$, minimum $-\sqrt{2}$ at $\left(-\sqrt{2}, 1\right)$ and $\left(\sqrt{2}, -1\right)$.

7. $12x^2 = 4x\lambda, 2y = 2y\lambda$. If $y \neq 0$ then $\lambda = 1$ and $12x^2 = 4x, 12x(x - 1/3) = 0, x = 0$ or $x = 1/3$ so from $2x^2 + y^2 = 1$ we find that $y = \pm 1$ when $x = 0, y = \pm\sqrt{7}/3$ when $x = 1/3$. If $y = 0$ then $2x^2 + (0)^2 = 1, x = \pm 1/\sqrt{2}$. Test $(0, \pm 1), \left(1/3, \pm\sqrt{7}/3\right)$, and $(\pm 1/\sqrt{2}, 0)$. $f(0, \pm 1) = 1$, $f\left(1/3, \pm\sqrt{7}/3\right) = 25/27, f\left(1/\sqrt{2}, 0\right) = \sqrt{2}, f\left(-1/\sqrt{2}, 0\right) = -\sqrt{2}$. Maximum $\sqrt{2}$ at $\left(1/\sqrt{2}, 0\right)$, minimum $-\sqrt{2}$ at $\left(-1/\sqrt{2}, 0\right)$.

9. $2 = 2x\lambda, 1 = 2y\lambda, -2 = 2z\lambda; 1/x = 1/(2y) = -1/z$ thus $x = 2y, z = -2y$ so $(2y)^2 + y^2 + (-2y)^2 = 4, y^2 = 4/9, y = \pm 2/3$. Test $(-4/3, -2/3, 4/3)$ and $(4/3, 2/3, -4/3)$. $f(-4/3, -2/3, 4/3) = -6, f(4/3, 2/3, -4/3) = 6$. Maximum 6 at $(4/3, 2/3, -4/3)$, minimum -6 at $(-4/3, -2/3, 4/3)$.

11. $yz = 2x\lambda, xz = 2y\lambda, xy = 2z\lambda; yz/(2x) = xz/(2y) = xy/(2z)$ thus $y^2 = x^2, z^2 = x^2$ so $x^2 + x^2 + x^2 = 1, x = \pm 1/\sqrt{3}$. Test the eight possibilities with $x = \pm 1/\sqrt{3}, y = \pm 1/\sqrt{3}$, and $z = \pm 1/\sqrt{3}$ to find the maximum is $1/\left(3\sqrt{3}\right)$ at $\left(1/\sqrt{3}, 1/\sqrt{3}, 1/\sqrt{3}\right), \left(1/\sqrt{3}, -1/\sqrt{3}, -1/\sqrt{3}\right)$, $\left(-1/\sqrt{3}, 1/\sqrt{3}, -1/\sqrt{3}\right)$, and $\left(-1/\sqrt{3}, -1/\sqrt{3}, 1/\sqrt{3}\right)$; the minimum is $-1/\left(3\sqrt{3}\right)$ at $\left(1/\sqrt{3}, 1/\sqrt{3}, -1/\sqrt{3}\right), \left(1/\sqrt{3}, -1/\sqrt{3}, 1/\sqrt{3}\right), \left(-1/\sqrt{3}, 1/\sqrt{3}, 1/\sqrt{3}\right)$, and $\left(-1/\sqrt{3}, -1/\sqrt{3}, -1/\sqrt{3}\right)$.

13. $f(x, y) = x^2 + y^2; 2x = 2\lambda, 2y = -4\lambda; y = -2x$ so $2x - 4(-2x) = 3, x = 3/10$. The point is $(3/10, -3/5)$.

15. $f(x, y, z) = x^2 + y^2 + z^2$; $2x = \lambda$, $2y = 2\lambda$, $2z = \lambda$; $y = 2x$, $z = x$ so $x + 2(2x) + x = 1$, $x = 1/6$. The point is $(1/6, 1/3, 1/6)$.

17. $f(x, y) = (x - 1)^2 + (y - 2)^2$; $2(x - 1) = 2x\lambda$, $2(y - 2) = 2y\lambda$; $(x - 1)/x = (y - 2)/y$, $y = 2x$ so $x^2 + (2x)^2 = 45$, $x = \pm 3$. $f(-3, -6) = 80$ and $f(3, 6) = 20$ so $(3, 6)$ is closest and $(-3, -6)$ is farthest.

19. $f(x, y, z) = x + y + z$, $x^2 + y^2 + z^2 = 25$ where x, y, and z are the components of the vector; $1 = 2x\lambda$, $1 = 2y\lambda$, $1 = 2z\lambda$; $1/(2x) = 1/(2y) = 1/(2z)$; $y = x$, $z = x$ so $x^2 + x^2 + x^2 = 25$, $x = \pm 5/\sqrt{3}$. $f\left(-5/\sqrt{3}, -5/\sqrt{3}, -5/\sqrt{3}\right) = -5\sqrt{3}$ and $f\left(5/\sqrt{3}, 5/\sqrt{3}, 5/\sqrt{3}\right) = 5\sqrt{3}$ so the vector is $5(\mathbf{i} + \mathbf{j} + \mathbf{k})/\sqrt{3}$.

21. Minimize $f = x^2 + y^2 + z^2$ subject to $g(x, y, z) = x + y + z - 27 = 0$. $\nabla f = \lambda \nabla g$, $2x\mathbf{i} + 2y\mathbf{j} + 2z\mathbf{k} = \lambda\mathbf{i} + \lambda\mathbf{j} + \lambda\mathbf{k}$, solution $x = y = z = 9$, minimum value 243

23. Minimize $f = x^2 + y^2 + z^2$ subject to $x^2 - yz = 5$, $\nabla f = \lambda \nabla g$, $2x = 2x\lambda$, $2y = -z\lambda$, $2z = -y\lambda$. If $\lambda \neq \pm 2$, then $y = z = 0$, $x = \pm\sqrt{5}$, $f = 5$; if $\lambda = \pm 2$ then $x = 0$, and since $-yz = 5$, $y = -z = \pm\sqrt{5}$, $f = 10$, thus the minimum value is 5 at $(\pm\sqrt{5}, 0, 0)$.

25. Let x, y, and z be, respectively, the length, width, and height of the box. Minimize $f(x, y, z) = 10(2xy) + 5(2xz + 2yz) = 10(2xy + xz + yz)$ subject to $g(x, y, z) = xyz - 16 = 0$, $\nabla f = \lambda \nabla g$, $20y + 10z = \lambda yz$, $20x + 10z = \lambda xz$, $10x + 10y = \lambda xy$. Since $V = xyz = 16$, $x, y, z \neq 0$, thus $\lambda z = 20 + 10(z/y) = 20 + 10(z/x)$, so $x = y$. From this and $10x + 10y = \lambda xy$ it follows that $20 = \lambda x$, so $10z = 20x$, $z = 2x = 2y$, $V = 2x^3 = 16$ and thus $x = y = 2$ ft, $z = 4$ ft, $f(2, 2, 4) = 240$ cents.

27. Maximize $A(a, b, \alpha) = ab\sin\alpha$ subject to $g(a, b, \alpha) = 2a + 2b - \ell = 0$, $\nabla_{(a,b,\alpha)}f = \lambda\nabla_{(a,b,\alpha)}g$, $b\sin\alpha = 2\lambda$, $a\sin\alpha = 2\lambda$, $ab\cos\alpha = 0$ with solution $a = b \ (= \ell/4)$, $\alpha = \pi/2$ maximum value if parallelogram is a square.

29. **(a)** Maximize $f(\alpha, \beta, \gamma) = \cos\alpha\cos\beta\cos\gamma$ subject to $g(\alpha, \beta, \gamma) = \alpha + \beta + \gamma - \pi = 0$, $\nabla f = \lambda\nabla g$, $-\sin\alpha\cos\beta\cos\gamma = \lambda$, $-\cos\alpha\sin\beta\cos\gamma = \lambda$, $-\cos\alpha\cos\beta\sin\gamma = \lambda$ with solution $\alpha = \beta = \gamma = \pi/3$, maximum value $1/8$

(b) for example, $f(\alpha, \beta) = \cos\alpha\cos\beta\cos(\pi - \alpha - \beta)$

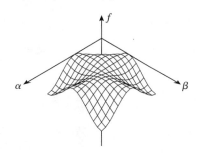

REVIEW EXERCISES, CHAPTER 14

1. **(a)** $f(\ln y, e^x) = e^{\ln y}\ln e^x = xy$ **(b)** $e^{r+s}\ln(rs)$

3. $z = \sqrt{x^2 + y^2} = c$ implies $x^2 + y^2 = c^2$, which is the equation of a circle; $x^2 + y^2 = c$ is also the equation of a circle (for $c > 0$).

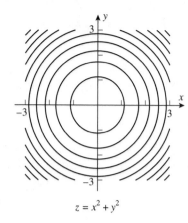

$$z = x^2 + y^2$$

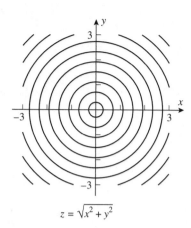

$$z = \sqrt{x^2 + y^2}$$

5. $x^4 - x + y - x^3 y = (x^3 - 1)(x - y)$, limit $= -1$, not defined on the line $y = x$ so not continuous at $(0,0)$

7. **(a)** They approximate the profit per unit of any additional sales of the standard or high-resolution monitors, respectively.

   (b) The rates of change with respect to the two directions x and y, and with respect to time.

9. **(a)** $P = \dfrac{10T}{V}$,

$$\frac{dP}{dt} = \frac{\partial P}{\partial T}\frac{dT}{dt} + \frac{\partial P}{\partial V}\frac{dV}{dt} = \frac{10}{V} \cdot 3 - \frac{10T}{V^2} \cdot 0 = \frac{30}{V} = \frac{30}{2.5} = 12 \text{ N/(m}^2\text{min)} = 12 \text{ Pa/min}$$

   (b) $\dfrac{dP}{dt} = \dfrac{\partial P}{\partial T}\dfrac{dT}{dt} + \dfrac{\partial P}{\partial V}\dfrac{dV}{dt} = \dfrac{10}{V} \cdot 0 - \dfrac{10T}{V^2} \cdot (-3) = \dfrac{30T}{V^2} = \dfrac{30 \cdot 50}{(2.5)^2} = 240 \text{ Pa/min}$

11. $w_x = 2x \sec^2(x^2 + y^2) + \sqrt{y},\ w_{xy} = 8xy \sec^2(x^2 + y^2)\tan(x^2 + y^2) + \dfrac{1}{2}y^{-1/2}$,

$w_y = 2y \sec^2(x^2 + y^2) + \dfrac{1}{2}xy^{-1/2},\ w_{yx} = 8xy \sec^2(x^2 + y^2)\tan(x^2 + y^2) + \dfrac{1}{2}y^{-1/2}$

13. $F_x = -6xz, F_{xx} = -6z, F_y = -6yz, F_{yy} = -6z, F_z = 6z^2 - 3x^2 - 3y^2$,

$F_{zz} = 12z, F_{xx} + F_{yy} + F_{zz} = -6z - 6z + 12z = 0$

17. $dV = \dfrac{2}{3}xh\,dx + \dfrac{1}{3}x^2\,dh = \dfrac{2}{3}2(-0.1) + \dfrac{1}{3}(0.2) = -0.06667 \text{ m}^3;\ \Delta V = -0.07267 \text{ m}^3$

19. $\dfrac{dz}{dt} = \dfrac{\partial z}{\partial x}\dfrac{dx}{dt} + \dfrac{\partial z}{\partial y}\dfrac{dy}{dt}$, so when $t = 0$, $4\left(-\dfrac{1}{2}\right) + 2\dfrac{dy}{dt} = 2$. Solve to obtain $\dfrac{dy}{dt}\bigg|_{t=0} = 2$

21. $\dfrac{dy}{dx} = -\dfrac{f_x}{f_y},\ \dfrac{d^2y}{dx^2} = -\dfrac{f_y(d/dx)f_x - f_x(d/dx)f_y}{f_y^2} = -\dfrac{f_y(f_{xx} + f_{xy}(dy/dx)) - f_x(f_{xy} + f_{yy}(dy/dx))}{f_y^2}$

$= -\dfrac{f_y(f_{xx} + f_{xy}(-f_x/f_y)) - f_x(f_{xy} + f_{yy}(-f_x/f_y))}{f_y^2} = \dfrac{-f_y^2 f_{xx} + 2f_x f_y f_{xy} - f_x^2 f_{yy}}{f_y^3}$

25. $\nabla f = \dfrac{y}{x + y}\mathbf{i} + \left(\ln(x + y) + \dfrac{y}{x + y}\right)\mathbf{j}$, so when $(x, y) = (-3, 5)$,

$\dfrac{\partial f}{\partial u} = \nabla f \cdot \mathbf{u} = \left[\dfrac{5}{2}\mathbf{i} + \left(\ln 2 + \dfrac{5}{2}\right)\mathbf{j}\right] \cdot \left[\dfrac{3}{5}\mathbf{i} + \dfrac{4}{5}\mathbf{j}\right] = \dfrac{3}{2} + 2 + \dfrac{4}{5}\ln 2 = \dfrac{7}{2} + \dfrac{4}{5}\ln 2$

27. Use the unit vectors $\mathbf{u} = \langle \frac{1}{\sqrt{2}}, \frac{1}{\sqrt{2}} \rangle, \mathbf{v} = \langle 0, -1 \rangle, \mathbf{w} = \langle -\frac{1}{\sqrt{5}}, -\frac{2}{\sqrt{5}} \rangle = -\frac{\sqrt{2}}{\sqrt{5}}\mathbf{u} + \frac{1}{\sqrt{5}}\mathbf{v}$, so that

$$D_{\mathbf{w}}f = -\frac{\sqrt{2}}{\sqrt{5}}D_{\mathbf{u}}f + \frac{1}{\sqrt{5}}D_{\mathbf{v}}f = -\frac{\sqrt{2}}{\sqrt{5}}2\sqrt{2} + \frac{1}{\sqrt{5}}(-3) = -\frac{7}{\sqrt{5}}$$

29. The origin is not such a point, so assume that the normal line at $(x_0, y_0, z_0) \neq (0, 0, 0)$ passes through the origin, then $\mathbf{n} = z_x\mathbf{i} + z_y\mathbf{j} - \mathbf{k} = -y_0\mathbf{i} - x_0\mathbf{j} - \mathbf{k}$; the line passes through the origin and is normal to the surface if it has the form $\mathbf{r}(t) = -y_0t\mathbf{i} - x_0t\mathbf{j} - t\mathbf{k}$ and $(x_0, y_0, z_0) = (x_0, y_0, 2 - x_0y_0)$ lies on the line if $-y_0t = x_0, -x_0t = y_0, -t = 2 - x_0y_0$, with solutions $x_0 = y_0 = -1$, $x_0 = y_0 = 1, x_0 = y_0 = 0$; thus the points are $(0, 0, 2), (1, 1, 1), (-1, -1, 1)$.

31. A tangent to the line is $6\mathbf{i} + 4\mathbf{j} + \mathbf{k}$, a normal to the surface is $\mathbf{n} = 18x\mathbf{i} + 8y\mathbf{j} - \mathbf{k}$, so solve

$18x = 6k, 8y = 4k, -1 = k; \quad k = -1, x = -1/3, y = -1/2, z = 2$

33. $\nabla f = (2x + 3y - 6)\mathbf{i} + (3x + 6y + 3)\mathbf{j} = \mathbf{0}$ if $2x + 3y = 6, x + 2y = -1, x = 15, y = -8, D = 3 > 0$, $f_{xx} = 2 > 0$, so f has a relative minimum at $(15, -8)$.

35. $\nabla f = (3x^2 - 3y)\mathbf{i} - (3x - y)\mathbf{j} = \mathbf{0}$ if $y = x^2, 3x = y$, so $x = y = 0$ or $x = 3, y = 9$; at $x = y = 0, D = -9$, saddle point; at $x = 3, y = 9, D = 9, f_{xx} = 18 > 0$, relative minimum

37. **(a)** $y^2 = 8 - 4x^2$, find extrema of $f(x) = x^2(8 - 4x^2) = -4x^4 + 8x^2$ defined for $-\sqrt{2} \leq x \leq \sqrt{2}$. Then $f'(x) = -16x^3 + 16x = 0$ when $x = 0, \pm1, f''(x) = -48x^2 + 16$, so f has a relative maximum at $x = \pm1, y = \pm2$ and a relative minimum at $x = 0, y = \pm2\sqrt{2}$. At the endpoints $x = \pm\sqrt{2}, y = 0$ we obtain the minimum $f = 0$ again.

(b) $f(x, y) = x^2y^2, g(x, y) = 4x^2 + y^2 - 8 = 0, \nabla f = 2xy^2\mathbf{i} + 2x^2y\mathbf{j} = \lambda\nabla g = 8\lambda x\mathbf{i} + 2\lambda y\mathbf{j}$, so solve $2xy^2 = \lambda 8x, 2x^2y = \lambda 2y$. If $x = 0$ then $y = \pm2\sqrt{2}$, and if $y = 0$ then $x = \pm\sqrt{2}$. In either case f has a relative and absolute minimum. Assume $x, y \neq 0$, then $y^2 = 4\lambda, x^2 = \lambda$, use $g = 0$ to obtain $x^2 = 1, x = \pm1, y = \pm2$, and $f = 4$ is a relative and absolute maximum at $(\pm1, \pm2)$.

39. Denote the currents I_1, I_2, I_3 by x, y, z respectively. Then minimize $F(x, y, z) = x^2R_1 + y^2R_2 + z^2R_3$ subject to $g(x, y, z) = x + y + z - I = 0$, so solve $\nabla F = \lambda\nabla g, 2xR_1\mathbf{i} + 2yR_2\mathbf{j} + 2zR_3\mathbf{k} = \lambda(\mathbf{i} + \mathbf{j} + \mathbf{k})$, $\lambda = 2xR_1 = 2yR_2 = 2zR_3$, so the minimum value of F occurs when $I_1 : I_2 : I_3 = \frac{1}{R_1} : \frac{1}{R_2} : \frac{1}{R_3}$.

41. **(a)** $\partial P/\partial L = c\alpha L^{\alpha-1}K^\beta, \partial P/\partial K = c\beta L^\alpha K^{\beta-1}$

(b) the rates of change of output with respect to labor and capital equipment, respectively

(c) $K(\partial P/\partial K) + L(\partial P/\partial L) = c\beta L^\alpha K^\beta + c\alpha L^\alpha K^\beta = (\alpha + \beta)P = P$

CHAPTER 15
Multiple Integrals

EXERCISE SET 15.1

1. $\int_0^1 \int_0^2 (x+3)dy\,dx = \int_0^1 (2x+6)dx = 7$ **3.** $\int_2^4 \int_0^1 x^2 y\,dx\,dy = \int_2^4 \frac{1}{3}y\,dy = 2$

5. $\int_0^{\ln 3} \int_0^{\ln 2} e^{x+y}dy\,dx = \int_0^{\ln 3} e^x dx = 2$ **7.** $\int_{-1}^0 \int_2^5 dx\,dy = \int_{-1}^0 3\,dy = 3$

9. $\int_0^1 \int_0^1 \frac{x}{(xy+1)^2}dy\,dx = \int_0^1 \left(1 - \frac{1}{x+1}\right)dx = 1 - \ln 2$

11. $\int_0^{\ln 2} \int_0^1 xy\,e^{y^2 x}dy\,dx = \int_0^{\ln 2} \frac{1}{2}(e^x - 1)dx = (1 - \ln 2)/2$

13. $\int_{-1}^1 \int_{-2}^2 4xy^3 dy\,dx = \int_{-1}^1 0\,dx = 0$

15. $\int_0^1 \int_2^3 x\sqrt{1-x^2}\,dy\,dx = \int_0^1 x(1-x^2)^{1/2}dx = 1/3$

17. **(a)** $x_k^* = k/2 - 1/4, k = 1,2,3,4; y_l^* = l/2 - 1/4, l = 1,2,3,4,$

$$\int\int_R f(x,y)\,dxdy \approx \sum_{k=1}^4 \sum_{l=1}^4 f(x_k^*, y_l^*)\Delta A_{kl} = \sum_{k=1}^4 \sum_{l=1}^4 [(k/2-1/4)^2 + (l/2-1/4)](1/2)^2 = 37/4$$

(b) $\int_0^2 \int_0^2 (x^2 + y)\,dxdy = 28/3$; the error is $|37/4 - 28/3| = 1/12$

19. **(a)** 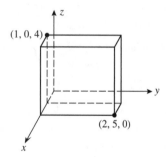   **(b)**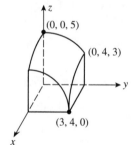

21. $V = \int_3^5 \int_1^2 (2x+y)dy\,dx = \int_3^5 (2x + 3/2)dx = 19$

23. $V = \int_0^2 \int_0^3 x^2 dy\,dx = \int_0^2 3x^2 dx = 8$

25. $\int_0^{1/2} \int_0^\pi x\cos(xy)\cos^2 \pi x\,dy\,dx = \int_0^{1/2} \cos^2 \pi x \sin(xy)\Big]_0^\pi dx$

$$= \int_0^{1/2} \cos^2 \pi x \sin \pi x\,dx = -\frac{1}{3\pi}\cos^3 \pi x\Big]_0^{1/2} = \frac{1}{3\pi}$$

27. $f_{\text{ave}} = \dfrac{2}{\pi} \displaystyle\int_0^{\pi/2} \int_0^1 y \sin xy \, dx \, dy = \dfrac{2}{\pi} \int_0^{\pi/2} \left(-\cos xy \right]_{x=0}^{x=1} \right) dy = \dfrac{2}{\pi} \int_0^{\pi/2} (1 - \cos y) \, dy = 1 - \dfrac{2}{\pi}$

29. $T_{\text{ave}} = \dfrac{1}{2} \displaystyle\int_0^1 \int_0^2 (10 - 8x^2 - 2y^2) \, dy \, dx = \dfrac{1}{2} \int_0^1 \left(\dfrac{44}{3} - 16x^2 \right) dx = \left(\dfrac{14}{3} \right)^{\circ} C$

31. 1.381737122

33. $\displaystyle\iint_R f(x,y) dA = \int_a^b \left[\int_c^d g(x) h(y) dy \right] dx = \int_a^b g(x) \left[\int_c^d h(y) dy \right] dx$

$$= \left[\int_a^b g(x) dx \right] \left[\int_c^d h(y) dy \right]$$

35. The first integral equals $1/2$, the second equals $-1/2$. No, because the integrand is not continuous.

EXERCISE SET 15.2

1. $\displaystyle\int_0^1 \int_{x^2}^x xy^2 \, dy \, dx = \int_0^1 \dfrac{1}{3}(x^4 - x^7) dx = 1/40$

3. $\displaystyle\int_0^3 \int_0^{\sqrt{9-y^2}} y \, dx \, dy = \int_0^3 y\sqrt{9 - y^2} \, dy = 9$

5. $\displaystyle\int_{\sqrt{\pi}}^{\sqrt{2\pi}} \int_0^{x^3} \sin(y/x) dy \, dx = \int_{\sqrt{\pi}}^{\sqrt{2\pi}} [-x\cos(x^2) + x] dx = \pi/2$

7. $\displaystyle\int_{\pi/2}^{\pi} \int_0^{x^2} \dfrac{1}{x} \cos(y/x) dy \, dx = \int_{\pi/2}^{\pi} \sin x \, dx = 1$

9. $\displaystyle\int_0^1 \int_0^x y\sqrt{x^2 - y^2} \, dy \, dx = \int_0^1 \dfrac{1}{3} x^3 dx = 1/12$

11. (a) $\displaystyle\int_0^2 \int_0^{x^2} f(x,y) \, dy dx$ (b) $\displaystyle\int_0^4 \int_{\sqrt{y}}^2 f(x,y) \, dx dy$

13. (a) $\displaystyle\int_1^2 \int_{-2x+5}^3 f(x,y) \, dy dx + \int_2^4 \int_1^3 f(x,y) \, dy dx + \int_4^5 \int_{2x-7}^3 f(x,y) \, dy dx$

(b) $\displaystyle\int_1^3 \int_{(5-y)/2}^{(y+7)/2} f(x,y) \, dx dy$

15. (a) $\displaystyle\int_0^2 \int_0^{x^2} xy \, dy \, dx = \int_0^2 \dfrac{1}{2} x^5 dx = \dfrac{16}{3}$

(b) $\displaystyle\int_1^3 \int_{-(y-5)/2}^{(y+7)/2} xy \, dx \, dy = \int_1^3 (3y^2 + 3y) dy = 38$

17. (a) $\displaystyle\int_4^8 \int_{16/x}^x x^2 \, dy \, dx = \int_4^8 (x^3 - 16x) dx = 576$

(b) $\displaystyle\int_2^4\int_{16/y}^8 x^2\,dxdy + \int_4^8\int_y^8 x^2\,dx\,dy = \int_4^8\left[\frac{512}{3} - \frac{4096}{3y^3}\right]dy + \int_4^8 \frac{512 - y^3}{3}\,dy$

$$= \frac{640}{3} + \frac{1088}{3} = 576$$

19. **(a)** $\displaystyle\int_{-1}^1\int_{-\sqrt{1-x^2}}^{\sqrt{1-x^2}}(3x - 2y)dy\,dx = \int_{-1}^1 6x\sqrt{1-x^2}\,dx = 0$

(b) $\displaystyle\int_{-1}^1\int_{-\sqrt{1-y^2}}^{\sqrt{1-y^2}}(3x - 2y)\,dxdy = \int_{-1}^1 -4y\sqrt{1-y^2}\,dy = 0$

21. $\displaystyle\int_0^4\int_0^{\sqrt{y}} x(1 + y^2)^{-1/2}dx\,dy = \int_0^4 \frac{1}{2}y(1 + y^2)^{-1/2}dy = (\sqrt{17} - 1)/2$

23. $\displaystyle\int_0^2\int_{y^2}^{6-y} xy\,dx\,dy = \int_0^2 \frac{1}{2}(36y - 12y^2 + y^3 - y^5)dy = 50/3$

25. $\displaystyle\int_0^1\int_{x^3}^x (x - 1)dy\,dx = \int_0^1(-x^4 + x^3 + x^2 - x)dx = -7/60$

27. **(a)**

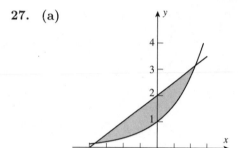

(b) $x = (-1.8414, 0.1586), (1.1462, 3.1462)$

(c) $\displaystyle\iint_R x\,dA \approx \int_{-1.8414}^{1.1462}\int_{e^x}^{x+2} x\,dydx = \int_{-1.8414}^{1.1462} x(x + 2 - e^x)\,dx \approx -0.4044$

(d) $\displaystyle\iint_R x\,dA \approx \int_{0.1586}^{3.1462}\int_{y-2}^{\ln y} x\,dxdy = \int_{0.1586}^{3.1462}\left[\frac{\ln^2 y}{2} - \frac{(y-2)^2}{2}\right]dy \approx -0.4044$

29. $A = \displaystyle\int_0^{\pi/4}\int_{\sin x}^{\cos x} dy\,dx = \int_0^{\pi/4}(\cos x - \sin x)dx = \sqrt{2} - 1$

31. $A = \displaystyle\int_{-3}^3\int_{1-y^2/9}^{9-y^2} dx\,dy = \int_{-3}^3 8(1 - y^2/9)dy = 32$

33. $\displaystyle\int_0^4\int_0^{6-3x/2}(3 - 3x/4 - y/2)\,dy\,dx = \int_0^4[(3 - 3x/4)(6 - 3x/2) - (6 - 3x/2)^2/4]\,dx = 12$

35. $V = \displaystyle\int_{-3}^{3} \int_{-\sqrt{9-x^2}}^{\sqrt{9-x^2}} (3-x)\,dy\,dx = \int_{-3}^{3}(6\sqrt{9-x^2} - 2x\sqrt{9-x^2})\,dx = 27\pi$

37. $V = \displaystyle\int_{0}^{3} \int_{0}^{2} (9x^2 + y^2)\,dy\,dx = \int_{0}^{3}(18x^2 + 8/3)\,dx = 170$

39. $V = \displaystyle\int_{-3/2}^{3/2} \int_{-\sqrt{9-4x^2}}^{\sqrt{9-4x^2}} (y+3)\,dy\,dx = \int_{-3/2}^{3/2} 6\sqrt{9-4x^2}\,dx = 27\pi/2$

41. $V = 8\displaystyle\int_{0}^{5} \int_{0}^{\sqrt{25-x^2}} \sqrt{25-x^2}\,dy\,dx = 8\int_{0}^{5}(25-x^2)\,dx = 2000/3$

43. $V = 4\displaystyle\int_{0}^{1} \int_{0}^{\sqrt{1-x^2}} (1-x^2-y^2)\,dy\,dx = \dfrac{8}{3}\int_{0}^{1}(1-x^2)^{3/2}\,dx = \pi/2$

45. $\displaystyle\int_{0}^{\sqrt{2}} \int_{y^2}^{2} f(x,y)\,dx\,dy$ **47.** $\displaystyle\int_{1}^{e^2} \int_{\ln x}^{2} f(x,y)\,dy\,dx$ **49.** $\displaystyle\int_{0}^{\pi/2} \int_{0}^{\sin x} f(x,y)\,dy\,dx$

51. $\displaystyle\int_{0}^{4} \int_{0}^{y/4} e^{-y^2}\,dx\,dy = \int_{0}^{4} \dfrac{1}{4} y e^{-y^2}\,dy = (1 - e^{-16})/8$

53. $\displaystyle\int_{0}^{2} \int_{0}^{x^2} e^{x^3}\,dy\,dx = \int_{0}^{2} x^2 e^{x^3}\,dx = (e^8 - 1)/3$

55. $\displaystyle\int_{0}^{2} \int_{0}^{y^2} \sin(y^3)\,dx\,dy = \int_{0}^{2} y^2 \sin(y^3)\,dy = (1 - \cos 8)/3$

57. (a) $\displaystyle\int_{0}^{4} \int_{\sqrt{x}}^{2} \sin \pi y^3 \, dy\,dx$; the inner integral is non-elementary.

$\displaystyle\int_{0}^{2} \int_{0}^{y^2} \sin\left(\pi y^3\right)\,dx\,dy = \int_{0}^{2} y^2 \sin\left(\pi y^3\right)\,dy = -\dfrac{1}{3\pi}\cos\left(\pi y^3\right)\Big]_{0}^{2} = 0$

(b) $\displaystyle\int_{0}^{1} \int_{\sin^{-1} y}^{\pi/2} \sec^2(\cos x)\,dx\,dy$; the inner integral is non-elementary.

$\displaystyle\int_{0}^{\pi/2} \int_{0}^{\sin x} \sec^2(\cos x)\,dy\,dx = \int_{0}^{\pi/2} \sec^2(\cos x)\sin x\,dx = \tan 1$

59. The region is symmetric with respect to the y-axis, and the integrand is an odd function of x, hence the answer is zero.

61. Area of triangle is $1/2$, so $\bar{f} = 2\displaystyle\int_{0}^{1} \int_{x}^{1} \dfrac{1}{1+x^2}\,dy\,dx = 2\int_{0}^{1}\left[\dfrac{1}{1+x^2} - \dfrac{x}{1+x^2}\right]dx = \dfrac{\pi}{2} - \ln 2$

63. $T_{\text{ave}} = \dfrac{1}{A(R)} \displaystyle\iint_{R} (5xy + x^2)\,dA$. The diamond has corners $(\pm 2, 0), (0, \pm 4)$ and thus has area

$A(R) = 4\dfrac{1}{2}2(4) = 16\text{m}^2$. Since $5xy$ is an odd function of x (as well as y), $\displaystyle\iint_{R} 5xy\,dA = 0$. Since

x^2 is an even function of both x and y,

$$T_{\text{ave}} = \frac{4}{16} \iint\limits_{\substack{R \\ x,y>0}} x^2 \, dA = \frac{1}{4} st_0^2 \int_0^{4-2x} x^2 \, dy dx = \frac{1}{4} \int_0^2 (4-2x)x^2 \, dx = \frac{1}{4}\left(\frac{4}{3}x^3 - \frac{1}{2}x^4\right)\Big]_0^2 = \frac{2}{3}{}^\circ\text{C}$$

65. $y = \sin x$ and $y = x/2$ intersect at $x = 0$ and $x = a = 1.895494$, so

$$V = \int_0^a \int_{x/2}^{\sin x} \sqrt{1+x+y} \, dy \, dx = 0.676089$$

EXERCISE SET 15.3

1. $\displaystyle\int_0^{\pi/2} \int_0^{\sin\theta} r\cos\theta dr \, d\theta = \int_0^{\pi/2} \frac{1}{2}\sin^2\theta\cos\theta \, d\theta = 1/6$

3. $\displaystyle\int_0^{\pi/2} \int_0^{a\sin\theta} r^2 dr \, d\theta = \int_0^{\pi/2} \frac{a^3}{3}\sin^3\theta \, d\theta = \frac{2}{9}a^3$

5. $\displaystyle\int_0^{\pi} \int_0^{1-\sin\theta} r^2\cos\theta \, dr \, d\theta = \int_0^{\pi} \frac{1}{3}(1-\sin\theta)^3\cos\theta \, d\theta = 0$

7. $\displaystyle A = \int_0^{2\pi} \int_0^{1-\cos\theta} r \, dr \, d\theta = \int_0^{2\pi} \frac{1}{2}(1-\cos\theta)^2 d\theta = 3\pi/2$

9. $\displaystyle A = \int_{\pi/4}^{\pi/2} \int_{\sin 2\theta}^{1} r \, dr \, d\theta = \int_{\pi/4}^{\pi/2} \frac{1}{2}(1-\sin^2 2\theta) d\theta = \pi/16$

11. $\displaystyle A = \int_{\pi/6}^{5\pi/6} \int_2^{4\sin\theta} f(r,\theta) \, r \, dr \, d\theta$ 　　　　**13.** $\displaystyle V = 8\int_0^{\pi/2} \int_1^3 r\sqrt{9-r^2} \, dr \, d\theta$

15. $\displaystyle V = 2\int_0^{\pi/2} \int_0^{\cos\theta} (1-r^2)r \, dr \, d\theta$

17. $\displaystyle V = 8\int_0^{\pi/2} \int_1^3 r\sqrt{9-r^2} \, dr \, d\theta = \frac{128}{3}\sqrt{2}\int_0^{\pi/2} d\theta = \frac{64}{3}\sqrt{2}\pi$

19. $\displaystyle V = 2\int_0^{\pi/2} \int_0^{\cos\theta} (1-r^2)r \, dr \, d\theta = \frac{1}{2}\int_0^{\pi/2} (2\cos^2\theta - \cos^4\theta) d\theta = 5\pi/32$

21. $\displaystyle V = \int_0^{\pi/2} \int_0^{3\sin\theta} r^2 \sin\theta \, dr d\theta = 9\int_0^{\pi/2} \sin^4\theta \, d\theta = \frac{27}{16}\pi$

23. $\displaystyle\int_0^{2\pi} \int_0^1 e^{-r^2} r \, dr \, d\theta = \frac{1}{2}(1-e^{-1})\int_0^{2\pi} d\theta = (1-e^{-1})\pi$

25. $\displaystyle\int_0^{\pi/4} \int_0^2 \frac{1}{1+r^2} r \, dr \, d\theta = \frac{1}{2}\ln 5\int_0^{\pi/4} d\theta = \frac{\pi}{8}\ln 5$

27. $\displaystyle\int_0^{\pi/2} \int_0^1 r^3 dr \, d\theta = \frac{1}{4}\int_0^{\pi/2} d\theta = \pi/8$

29. $\int_0^{\pi/2} \int_0^{2\cos\theta} r^2\, dr\, d\theta = \frac{8}{3}\int_0^{\pi/2} \cos^3\theta\, d\theta = 16/9$

31. $\int_0^{\pi/2} \int_0^a \frac{r}{(1+r^2)^{3/2}}\, dr\, d\theta = \frac{\pi}{2}\left(1 - 1/\sqrt{1+a^2}\right)$

33. $\int_0^{\pi/4} \int_0^2 \frac{r}{\sqrt{1+r^2}}\, dr\, d\theta = \frac{\pi}{4}(\sqrt{5}-1)$

35. $V = \int_0^{2\pi} \int_0^a hr\, dr\, d\theta = \int_0^{2\pi} h\frac{a^2}{2}\, d\theta = \pi a^2 h$

37. $V = 2\int_0^{\pi/2} \int_0^{a\sin\theta} \frac{c}{a}(a^2 - r^2)^{1/2} r\, dr\, d\theta = \frac{2}{3}a^2 c \int_0^{\pi/2}(1 - \cos^3\theta)d\theta = (3\pi - 4)a^2 c/9$

39. $A = \int_{\pi/6}^{\pi/4} \int_{\sqrt{8\cos 2\theta}}^{4\sin\theta} r\, dr\, d\theta + \int_{\pi/4}^{\pi/2} \int_0^{4\sin\theta} r\, dr\, d\theta$

$= \int_{\pi/6}^{\pi/4}(8\sin^2\theta - 4\cos 2\theta)d\theta + \int_{\pi/4}^{\pi/2} 8\sin^2\theta\, d\theta = 4\pi/3 + 2\sqrt{3} - 2$

41. (a) $I^2 = \left[\int_0^{+\infty} e^{-x^2}dx\right]\left[\int_0^{+\infty} e^{-y^2}dy\right] = \int_0^{+\infty}\left[\int_0^{+\infty} e^{-x^2}dx\right]e^{-y^2}dy$

$= \int_0^{+\infty} \int_0^{+\infty} e^{-x^2}e^{-y^2}dx\, dy = \int_0^{+\infty} \int_0^{+\infty} e^{-(x^2+y^2)}dx\, dy$

(b) $I^2 = \int_0^{\pi/2} \int_0^{+\infty} e^{-r^2} r\, dr\, d\theta = \frac{1}{2}\int_0^{\pi/2} d\theta = \pi/4$ **(c)** $I = \sqrt{\pi}/2$

43. (a) 1.173108605   **(b)** $\int_0^\pi \int_0^1 re^{-r^4}dr\, d\theta = \pi\int_0^1 re^{-r^4}\, dr \approx 1.173108605$

45. $\int_{\tan^{-1}(1/3)}^{\tan^{-1}(2)} \int_0^2 r^3\cos^2\theta\, dr\, d\theta = 4\int_{\tan^{-1}(1/3)}^{\tan^{-1}(2)} \cos^2\theta\, d\theta = 2\int_{\tan^{-1}(1/3)}^{\tan^{-1}(2)}(1 + \cos(2\theta))\, d\theta$

$= 2(\tan^{-1} 2 - \tan^{-1}(1/3)) + 2/\sqrt{5} - 1/\sqrt{10}$

EXERCISE SET 15.4

1. (a)

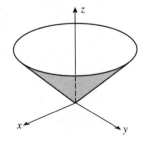

(b)

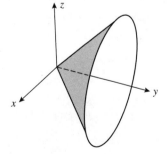

(c)

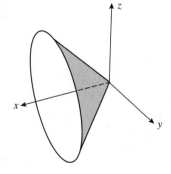

3. (a) $x = u, y = v, z = \dfrac{5}{2} + \dfrac{3}{2}u - 2v$ **(b)** $x = u, y = v, z = u^2$

5. (a) $x = 5\cos u, y = 5\sin u, z = v; 0 \le u \le 2\pi, 0 \le v \le 1$

    (b) $x = 2\cos u, y = v, z = 2\sin u; 0 \le u \le 2\pi, 1 \le v \le 3$

7. $x = u, y = \sin u \cos v, z = \sin u \sin v$ **9.** $x = r\cos\theta, y = r\sin\theta, z = \dfrac{1}{1 + r^2}$

11. $x = r\cos\theta, y = r\sin\theta, z = 2r^2\cos\theta\sin\theta$ **13.** $x = r\cos\theta, y = r\sin\theta, z = \sqrt{9 - r^2}; r \le \sqrt{5}$

15. $x = \dfrac{1}{2}\rho\cos\theta, y = \dfrac{1}{2}\rho\sin\theta, z = \dfrac{\sqrt{3}}{2}\rho$ **17.** $z = x - 2y$; a plane

19. $(x/3)^2 + (y/2)^2 = 1; 2 \le z \le 4$; part of an elliptic cylinder

21. $(x/3)^2 + (y/4)^2 = z^2; 0 \le z \le 1$; part of an elliptic cone

23. (a) $x = r\cos\theta, y = r\sin\theta, z = r, 0 \le r \le 2; x = u, y = v, z = \sqrt{u^2 + v^2}; 0 \le u^2 + v^2 \le 4$

25. (a) $0 \le u \le 3, 0 \le v \le \pi$ **(b)** $0 \le u \le 4, -\pi/2 \le v \le \pi/2$

27. (a) $0 \le \phi \le \pi/2, 0 \le \theta \le 2\pi$ **(b)** $0 \le \phi \le \pi, 0 \le \theta \le \pi$

29. $u = 1, v = 2, \mathbf{r}_u \times \mathbf{r}_v = -2\mathbf{i} - 4\mathbf{j} + \mathbf{k}; 2x + 4y - z = 5$

31. $u = 0, v = 1, \mathbf{r}_u \times \mathbf{r}_v = 6\mathbf{k}; z = 0$

33. $\mathbf{r}_u \times \mathbf{r}_v = (\sqrt{2}/2)\mathbf{i} - (\sqrt{2}/2)\mathbf{j} + (1/2)\mathbf{k}; x - y + \dfrac{\sqrt{2}}{2}z = \dfrac{\pi\sqrt{2}}{8}$

35. $z = \sqrt{9 - y^2}, z_x = 0, z_y = -y/\sqrt{9 - y^2}, z_x^2 + z_y^2 + 1 = 9/(9 - y^2)$,

$$S = \int_0^2 \int_{-3}^3 \frac{3}{\sqrt{9 - y^2}}\, dy\, dx = \int_0^2 3\pi\, dx = 6\pi$$

37. $z^2 = 4x^2 + 4y^2, 2zz_x = 8x$ so $z_x = 4x/z$, similarly $z_y = 4y/z$ thus

$$z_x^2 + z_y^2 + 1 = (16x^2 + 16y^2)/z^2 + 1 = 5, \ S = \int_0^1 \int_{x^2}^x \sqrt{5}\, dy\, dx = \sqrt{5}\int_0^1 (x - x^2)dx = \sqrt{5}/6$$

39. $z_x = -2x, z_y = -2y, z_x^2 + z_y^2 + 1 = 4x^2 + 4y^2 + 1$,

$$S = \iint\limits_R \sqrt{4x^2 + 4y^2 + 1}\, dA = \int_0^{2\pi} \int_0^1 r\sqrt{4r^2 + 1}\, dr\, d\theta$$

$$= \frac{1}{12}(5\sqrt{5} - 1)\int_0^{2\pi} d\theta = (5\sqrt{5} - 1)\pi/6$$

41. $\partial\mathbf{r}/\partial u = \cos v\mathbf{i} + \sin v\mathbf{j} + 2u\mathbf{k}, \partial\mathbf{r}/\partial v = -u\sin v\mathbf{i} + u\cos v\mathbf{j}$,

$$\|\partial\mathbf{r}/\partial u \times \partial\mathbf{r}/\partial v\| = u\sqrt{4u^2 + 1}; \ S = \int_0^{2\pi} \int_1^2 u\sqrt{4u^2 + 1}\, du\, dv = (17\sqrt{17} - 5\sqrt{5})\pi/6$$

43. $z_x = y$, $z_y = x$, $z_x^2 + z_y^2 + 1 = x^2 + y^2 + 1$,

$$S = \iint\limits_R \sqrt{x^2 + y^2 + 1}\, dA = \int_0^{\pi/6} \int_0^3 r\sqrt{r^2 + 1}\, dr\, d\theta = \frac{1}{3}(10\sqrt{10} - 1) \int_0^{\pi/6} d\theta = (10\sqrt{10} - 1)\pi/18$$

45. On the sphere, $z_x = -x/z$ and $z_y = -y/z$ so $z_x^2 + z_y^2 + 1 = (x^2 + y^2 + z^2)/z^2 = 16/(16 - x^2 - y^2)$; the planes $z = 1$ and $z = 2$ intersect the sphere along the circles $x^2 + y^2 = 15$ and $x^2 + y^2 = 12$;

$$S = \iint\limits_R \frac{4}{\sqrt{16 - x^2 - y^2}}\, dA = \int_0^{2\pi} \int_{\sqrt{12}}^{\sqrt{15}} \frac{4r}{\sqrt{16 - r^2}}\, dr\, d\theta = 4\int_0^{2\pi} d\theta = 8\pi$$

47. $\mathbf{r}(u, v) = a\cos u \sin v\mathbf{i} + a\sin u \sin v\mathbf{j} + a\cos v\mathbf{k}$, $\|\mathbf{r}_u \times \mathbf{r}_v\| = a^2 \sin v$,

$$S = \int_0^{\pi} \int_0^{2\pi} a^2 \sin v\, du\, dv = 2\pi a^2 \int_0^{\pi} \sin v\, dv = 4\pi a^2$$

49. $z_x = \dfrac{h}{a}\dfrac{x}{\sqrt{x^2 + y^2}}$, $z_y = \dfrac{h}{a}\dfrac{y}{\sqrt{x^2 + y^2}}$, $z_x^2 + z_y^2 + 1 = \dfrac{h^2 x^2 + h^2 y^2}{a^2(x^2 + y^2)} + 1 = (a^2 + h^2)/a^2$,

$$S = \int_0^{2\pi} \int_0^a \frac{\sqrt{a^2 + h^2}}{a} r\, dr\, d\theta = \frac{1}{2}a\sqrt{a^2 + h^2} \int_0^{2\pi} d\theta = \pi a\sqrt{a^2 + h^2}$$

51. $\partial\mathbf{r}/\partial u = -(a + b\cos v)\sin u\mathbf{i} + (a + b\cos v)\cos u\mathbf{j}$,

$\partial\mathbf{r}/\partial v = -b\sin v\cos u\mathbf{i} - b\sin v\sin u\mathbf{j} + b\cos v\mathbf{k}$, $\|\partial\mathbf{r}/\partial u \times \partial\mathbf{r}/\partial v\| = b(a + b\cos v)$;

$$S = \int_0^{2\pi} \int_0^{2\pi} b(a + b\cos v)du\, dv = 4\pi^2 ab$$

53. $z = -1$ when $v \approx 0.27955$, $z = 1$ when $v \approx 2.86204$, $\|\mathbf{r}_u \times \mathbf{r}_v\| = |\cos v|$;

$$S = \int_0^{2\pi} \int_{0.27955}^{2.86204} |\cos v|\, dv\, du \approx 9.099$$

55. (a) $(x/a)^2 + (y/b)^2 + (z/c)^2 = \sin^2\phi\cos^2\theta + \sin^2\phi\sin^2\theta + \cos^2\phi = \sin^2\phi + \cos^2\phi = 1$, an ellipsoid

(b) $\mathbf{r}(\phi, \theta) = \langle 2\sin\phi\cos\theta, 3\sin\phi\sin\theta, 4\cos\phi\rangle$; $\mathbf{r}_\phi \times \mathbf{r}_\theta = 2\langle 6\sin^2\phi\cos\theta, 4\sin^2\phi\sin\theta, 3\cos\phi\sin\phi\rangle$,

$\|\mathbf{r}_\phi \times \mathbf{r}_\theta\| = 2\sqrt{16\sin^4\phi + 20\sin^4\phi\cos^2\theta + 9\sin^2\phi\cos^2\phi}$,

$$S = \int_0^{2\pi} \int_0^{\pi} 2\sqrt{16\sin^4\phi + 20\sin^4\phi\cos^2\theta + 9\sin^2\phi\cos^2\phi}\, d\phi\, d\theta \approx 111.5457699$$

57. $\left(\dfrac{x}{a}\right)^2 + \left(\dfrac{y}{b}\right)^2 + \left(\dfrac{z}{c}\right)^2 = 1$, ellipsoid

59. $-\left(\dfrac{x}{a}\right)^2 - \left(\dfrac{y}{b}\right)^2 + \left(\dfrac{z}{c}\right)^2 = 1$, hyperboloid of two sheets

EXERCISE SET 15.5

1. $\displaystyle\int_{-1}^1 \int_0^2 \int_0^1 (x^2 + y^2 + z^2)dx\, dy\, dz = \int_{-1}^1 \int_0^2 (1/3 + y^2 + z^2)dy\, dz = \int_{-1}^1 (10/3 + 2z^2)dz = 8$

3. $\displaystyle\int_0^2 \int_{-1}^{y^2} \int_{-1}^z yz\, dx\, dz\, dy = \int_0^2 \int_{-1}^{y^2} (yz^2 + yz)dz\, dy = \int_0^2 \left(\frac{1}{3}y^7 + \frac{1}{2}y^5 - \frac{1}{6}y\right)dy = \frac{47}{3}$

5. $\displaystyle\int_0^3\int_0^{\sqrt{9-z^2}}\int_0^x xy\,dy\,dx\,dz = \int_0^3\int_0^{\sqrt{9-z^2}}\frac{1}{2}x^3dx\,dz = \int_0^3\frac{1}{8}(81-18z^2+z^4)dz = 81/5$

7. $\displaystyle\int_0^2\int_0^{\sqrt{4-x^2}}\int_{-5+x^2+y^2}^{3-x^2-y^2}x\,dz\,dy\,dx = \int_0^2\int_0^{\sqrt{4-x^2}}[2x(4-x^2)-2xy^2]dy\,dx$

$$= \int_0^2\frac{4}{3}x(4-x^2)^{3/2}dx = 128/15$$

9. $\displaystyle\int_0^\pi\int_0^1\int_0^{\pi/6}xy\sin yz\,dz\,dy\,dx = \int_0^\pi\int_0^1 x[1-\cos(\pi y/6)]dy\,dx = \int_0^\pi(1-3/\pi)x\,dx = \pi(\pi-3)/2$

11. $\displaystyle\int_0^{\sqrt{2}}\int_0^x\int_0^{2-x^2}xyz\,dz\,dy\,dx = \int_0^{\sqrt{2}}\int_0^x\frac{1}{2}xy(2-x^2)^2dy\,dx = \int_0^{\sqrt{2}}\frac{1}{4}x^3(2-x^2)^2dx = 1/6$

13. $\displaystyle\int_0^3\int_1^2\int_{-2}^1\frac{\sqrt{x+z^2}}{y}dz\,dy\,dx \approx 9.425$

15. $\displaystyle V = \int_0^4\int_0^{(4-x)/2}\int_0^{(12-3x-6y)/4}dz\,dy\,dx = \int_0^4\int_0^{(4-x)/2}\frac{1}{4}(12-3x-6y)dy\,dx$

$$= \int_0^4\frac{3}{16}(4-x)^2dx = 4$$

17. $\displaystyle V = 2\int_0^2\int_{x^2}^4\int_0^{4-y}dz\,dy\,dx = 2\int_0^2\int_{x^2}^4(4-y)dy\,dx = 2\int_0^2\left(8-4x^2+\frac{1}{2}x^4\right)dx = 256/15$

19. The projection of the curve of intersection onto the xy-plane is $x^2+y^2 = 1$,

    (a) $\displaystyle V = \int_{-1}^1\int_{-\sqrt{1-x^2}}^{\sqrt{1-x^2}}\int_{4x^2+y^2}^{4-3y^2}f(x,y,z)dz\,dy\,dx$

    (b) $\displaystyle V = \int_{-1}^1\int_{-\sqrt{1-y^2}}^{\sqrt{1-y^2}}\int_{4x^2+y^2}^{4-3y^2}f(x,y,z)dz\,dx\,dy$

21. The projection of the curve of intersection onto the xy-plane is $x^2+y^2 = 1$,

$$V = 4\int_0^1\int_0^{\sqrt{1-x^2}}\int_{4x^2+y^2}^{4-3y^2}dz\,dy\,dx$$

23. $\displaystyle V = 2\int_{-3}^3\int_0^{\sqrt{9-x^2}/3}\int_0^{x+3}dz\,dy\,dx$

25. **(a)**

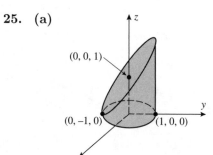

(b)

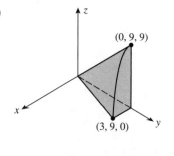

(c)

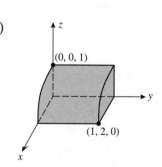

27. $V = \displaystyle\int_0^1 \int_0^{1-x} \int_0^{1-x-y} dz\,dy\,dx = 1/6, \quad f_{\text{ave}} = 6 \int_0^1 \int_0^{1-x} \int_0^{1-x-y} (x+y+z)\,dz\,dy\,dx = \dfrac{3}{4}$

29. The volume $V = \dfrac{3\pi}{\sqrt{2}}$, and thus

$$r_{\text{ave}} = \frac{\sqrt{2}}{3\pi} \iiint\limits_G \sqrt{x^2+y^2+z^2}\,dV$$

$$= \frac{\sqrt{2}}{3\pi} \int_{-1/\sqrt{2}}^{1/\sqrt{2}} \int_{-\sqrt{1-2x^2}}^{\sqrt{1-2x^2}} \int_{5x^2+5y^2}^{6-7x^2-y^2} \sqrt{x^2+y^2+z^2}\,dz\,dy\,dx \approx 3.291$$

31. (a) $\displaystyle\int_0^a \int_0^{b(1-x/a)} \int_0^{c(1-x/a-y/b)} dz\,dy\,dx, \quad \int_0^b \int_0^{a(1-y/b)} \int_0^{c(1-x/a-y/b)} dz\,dx\,dy,$

$\displaystyle\int_0^c \int_0^{a(1-z/c)} \int_0^{b(1-x/a-z/c)} dy\,dx\,dz, \quad \int_0^a \int_0^{c(1-x/a)} \int_0^{b(1-x/a-z/c)} dy\,dz\,dx,$

$\displaystyle\int_0^c \int_0^{b(1-z/c)} \int_0^{a(1-y/b-z/c)} dx\,dy\,dz, \quad \int_0^b \int_0^{c(1-y/b)} \int_0^{a(1-y/b-z/c)} dx\,dz\,dy$

(b) Use the first integral in Part (a) to get

$$\int_0^a \int_0^{b(1-x/a)} c\left(1 - \frac{x}{a} - \frac{y}{b}\right) dy\,dx = \int_0^a \frac{1}{2}bc\left(1 - \frac{x}{a}\right)^2 dx = \frac{1}{6}abc$$

33. (a) $\displaystyle\int_0^2 \int_0^{\sqrt{4-x^2}} \int_0^5 f(x,y,z)\,dz\,dy\,dx$

(b) $\displaystyle\int_0^9 \int_0^{3-\sqrt{x}} \int_y^{3-\sqrt{x}} f(x,y,z)\,dz\,dy\,dx$ **(c)** $\displaystyle\int_0^2 \int_0^{4-x^2} \int_y^{8-y} f(x,y,z)\,dz\,dy\,dx$

35. (a) At any point outside the closed sphere $\{x^2+y^2+z^2 \le 1\}$ the integrand is negative, so to maximize the integral it suffices to include all points inside the sphere; hence the maximum value is taken on the region $G = \{x^2+y^2+z^2 \le 1\}$.

(b) 4.934802202 **(c)** $\displaystyle\int_0^{2\pi} \int_0^\pi \int_0^1 (1-\rho^2)\rho\,d\rho\,d\phi\,d\theta = \frac{\pi^2}{2}$

37. (a) $\left[\displaystyle\int_{-1}^1 x\,dx\right] \left[\int_0^1 y^2\,dy\right] \left[\int_0^{\pi/2} \sin z\,dz\right] = (0)(1/3)(1) = 0$

(b) $\left[\displaystyle\int_0^1 e^{2x}\,dx\right] \left[\int_0^{\ln 3} e^y\,dy\right] \left[\int_0^{\ln 2} e^{-z}\,dz\right] = [(e^2-1)/2](2)(1/2) = (e^2-1)/2$

EXERCISE SET 15.6

1. (a) m_1 and m_3 are equidistant from $x = 5$, but m_3 has a greater mass, so the sum is positive.

(b) Let a be the unknown coordinate of the fulcrum; then the total moment about the fulcrum is $5(0-a) + 10(5-a) + 20(10-a) = 0$ for equilibrium, so $250 - 35a = 0$, $a = 50/7$. The fulcrum should be placed $50/7$ units to the right of m_1.

3. $A = 1, \quad \bar{x} = \displaystyle\int_0^1 \int_0^1 x\,dy\,dx = \frac{1}{2}, \quad \bar{y} = \int_0^1 \int_0^1 y\,dy\,dx = \frac{1}{2}$

5. $A = 1/2$, $\displaystyle\iint_R x\,dA = \int_0^1\int_0^x x\,dy\,dx = 1/3$, $\displaystyle\iint_R y\,dA = \int_0^1\int_0^x y\,dy\,dx = 1/6$;

centroid $(2/3, 1/3)$

7. $A = \displaystyle\int_0^1\int_x^{2-x^2} dy\,dx = 7/6$, $\displaystyle\iint_R x\,dA = \int_0^1\int_x^{2-x^2} x\,dy\,dx = 5/12$,

$\displaystyle\iint_R y\,dA = \int_0^1\int_x^{2-x^2} y\,dy\,dx = 19/15$; centroid $(5/14, 38/35)$

9. $\bar{x} = 0$ from the symmetry of the region,

$A = \dfrac{1}{2}\pi(b^2 - a^2)$, $\displaystyle\iint_R y\,dA = \int_0^\pi\int_a^b r^2\sin\theta\,dr\,d\theta = \dfrac{2}{3}(b^3 - a^3)$; centroid $\bar{x} = 0$, $\bar{y} = \dfrac{4(b^3 - a^3)}{3\pi(b^2 - a^2)}$.

11. $M = \displaystyle\iint_R \delta(x,y)dA = \int_0^1\int_0^1 |x + y - 1|\,dx\,dy$

$= \displaystyle\int_0^1\left[\int_0^{1-x}(1 - x - y)\,dy + \int_{1-x}^1 (x + y - 1)\,dy\right]dx = \dfrac{1}{3}$

$\bar{x} = 3\displaystyle\int_0^1\int_0^1 x\delta(x,y)\,dy\,dx = 3\int_0^1\left[\int_0^{1-x} x(1 - x - y)\,dy + \int_{1-x}^1 x(x + y - 1)\,dy\right]dx = \dfrac{1}{2}$

By symmetry, $\bar{y} = \dfrac{1}{2}$ as well; center of gravity $(1/2, 1/2)$

13. $M = \displaystyle\int_0^1\int_0^{\sqrt{x}}(x + y)dy\,dx = 13/20$, $M_x = \displaystyle\int_0^1\int_0^{\sqrt{x}}(x + y)y\,dy\,dx = 3/10$,

$M_y = \displaystyle\int_0^1\int_0^{\sqrt{x}}(x + y)x\,dy\,dx = 19/42$, $\bar{x} = M_y/M = 190/273$, $\bar{y} = M_x/M = 6/13$;

the mass is $13/20$ and the center of gravity is at $(190/273, 6/13)$.

15. $M = \displaystyle\int_0^{\pi/2}\int_0^a r^3\sin\theta\cos\theta\,dr\,d\theta = a^4/8$, $\bar{x} = \bar{y}$ from the symmetry of the density and the

region, $M_y = \displaystyle\int_0^{\pi/2}\int_0^a r^4\sin\theta\cos^2\theta\,dr\,d\theta = a^5/15$, $\bar{x} = 8a/15$; mass $a^4/8$, center of gravity $(8a/15, 8a/15)$.

17. $V = 1, \bar{x} = \displaystyle\int_0^1\int_0^1\int_0^1 x\,dz\,dy\,dx = \dfrac{1}{2}$, similarly $\bar{y} = \bar{z} = \dfrac{1}{2}$; centroid $\left(\dfrac{1}{2}, \dfrac{1}{2}, \dfrac{1}{2}\right)$

19. $\bar{x} = \bar{y} = \bar{z}$ from the symmetry of the region, $V = 1/6$,

$\bar{x} = \dfrac{1}{V}\displaystyle\int_0^1\int_0^{1-x}\int_0^{1-x-y} x\,dz\,dy\,dx = (6)(1/24) = 1/4$; centroid $(1/4, 1/4, 1/4)$

21. $\bar{x} = 1/2$ and $\bar{y} = 0$ from the symmetry of the region,

$V = \displaystyle\int_0^1\int_{-1}^1\int_{y^2}^1 dz\,dy\,dx = 4/3$, $\bar{z} = \dfrac{1}{V}\iiint_G z\,dV = (3/4)(4/5) = 3/5$; centroid $(1/2, 0, 3/5)$

23. $\bar{x} = \bar{y} = \bar{z}$ from the symmetry of the region, $V = \pi a^3/6$,

$$\bar{x} = \frac{1}{V} \int_0^a \int_0^{\sqrt{a^2-x^2}} \int_0^{\sqrt{a^2-x^2-y^2}} x \, dz \, dy \, dx = \frac{1}{V} \int_0^a \int_0^{\sqrt{a^2-x^2}} x\sqrt{a^2-x^2-y^2} \, dy \, dx$$

$$= \frac{1}{V} \int_0^{\pi/2} \int_0^a r^2 \sqrt{a^2-r^2} \cos\theta \, dr \, d\theta = \frac{6}{\pi a^3}(\pi a^4/16) = 3a/8; \text{ centroid } (3a/8, 3a/8, 3a/8)$$

25. $M = \int_0^a \int_0^a \int_0^a (a-x)dz \, dy \, dx = a^4/2, \, \bar{y} = \bar{z} = a/2$ from the symmetry of density and

region, $\bar{x} = \frac{1}{M} \int_0^a \int_0^a \int_0^a x(a-x)dz \, dy \, dx = (2/a^4)(a^5/6) = a/3$;

mass $a^4/2$, center of gravity $(a/3, a/2, a/2)$

27. $M = \int_{-1}^1 \int_0^1 \int_0^{1-y^2} yz \, dz \, dy \, dx = 1/6, \, \bar{x} = 0$ by the symmetry of density and region,

$$\bar{y} = \frac{1}{M} \iiint_G y^2 z \, dV = (6)(8/105) = 16/35, \, \bar{z} = \frac{1}{M} \iiint_G yz^2 dV = (6)(1/12) = 1/2;$$

mass $1/6$, center of gravity $(0, 16/35, 1/2)$

29. (a) $M = \int_0^1 \int_0^1 k(x^2+y^2)dy \, dx = 2k/3, \, \bar{x} = \bar{y}$ from the symmetry of density and region,

$$\bar{x} = \frac{1}{M} \iint_R kx(x^2+y^2)dA = \frac{3}{2k}(5k/12) = 5/8; \text{ center of gravity } (5/8, 5/8)$$

(b) $\bar{y} = 1/2$ from the symmetry of density and region,

$$M = \int_0^1 \int_0^1 kx \, dy \, dx = k/2, \, \bar{x} = \frac{1}{M} \iint_R kx^2 dA = (2/k)(k/3) = 2/3,$$

center of gravity $(2/3, 1/2)$

31. $V = \iiint_G dV = \int_0^\pi \int_0^{\sin x} \int_0^{1/(1+x^2+y^2)} dz \, dy \, dx = 0.666633$,

$$\bar{x} = \frac{1}{V} \iiint_G x dV = 1.177406, \, \bar{y} = \frac{1}{V} \iiint_G y dV = 0.353554, \, \bar{z} = \frac{1}{V} \iiint_G z dV = 0.231557$$

33. Let $x = r\cos\theta$, $y = r\sin\theta$, and $dA = r \, dr \, d\theta$ in formulas (11) and (12).

35. $\bar{x} = \bar{y}$ from the symmetry of the region, $A = \int_0^{\pi/2} \int_0^{\sin 2\theta} r \, dr \, d\theta = \pi/8$,

$$\bar{x} = \frac{1}{A} \int_0^{\pi/2} \int_0^{\sin 2\theta} r^2 \cos\theta \, dr \, d\theta = (8/\pi)(16/105) = \frac{128}{105\pi}; \text{ centroid } \left(\frac{128}{105\pi}, \frac{128}{105\pi}\right)$$

37. $\bar{x} = 0$ from the symmetry of the region, $\pi a^2/2$ is the area of the semicircle, $2\pi\bar{y}$ is the distance traveled by the centroid to generate the sphere so $4\pi a^3/3 = (\pi a^2/2)(2\pi\bar{y})$, $\bar{y} = 4a/(3\pi)$

39. $\bar{x} = k$ so $V = (\pi ab)(2\pi k) = 2\pi^2 abk$

41. The region generates a cone of volume $\frac{1}{3}\pi ab^2$ when it is revolved about the x-axis, the area of the region is $\frac{1}{2}ab$ so $\frac{1}{3}\pi ab^2 = \left(\frac{1}{2}ab\right)(2\pi\bar{y})$, $\bar{y} = b/3$. A cone of volume $\frac{1}{3}\pi a^2 b$ is generated when the region is revolved about the y-axis so $\frac{1}{3}\pi a^2 b = \left(\frac{1}{2}ab\right)(2\pi\bar{x})$, $\bar{x} = a/3$. The centroid is $(a/3, b/3)$.

43. $I_x = \displaystyle\int_0^a \int_0^b y^2 \delta\, dy\, dx = \frac{1}{3}\delta ab^3$, $I_y = \displaystyle\int_0^a \int_0^b x^2 \delta\, dy\, dx = \frac{1}{3}\delta a^3 b$,

$I_z = \displaystyle\int_0^a \int_0^b (x^2 + y^2)\delta\, dy\, dx = \frac{1}{3}\delta ab(a^2 + b^2)$

EXERCISE SET 15.7

1. $\displaystyle\int_0^{2\pi} \int_0^1 \int_0^{\sqrt{1-r^2}} zr\, dz\, dr\, d\theta = \int_0^{2\pi} \int_0^1 \frac{1}{2}(1-r^2)r\, dr\, d\theta = \int_0^{2\pi} \frac{1}{8}d\theta = \pi/4$

3. $\displaystyle\int_0^{\pi/2} \int_0^{\pi/2} \int_0^1 \rho^3 \sin\phi\cos\phi\, d\rho\, d\phi\, d\theta = \int_0^{\pi/2} \int_0^{\pi/2} \frac{1}{4}\sin\phi\cos\phi\, d\phi\, d\theta = \int_0^{\pi/2} \frac{1}{8}d\theta = \pi/16$

5. $f(r, \theta, z) = z$

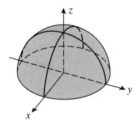

7. $f(\rho, \phi, \theta) = \rho\cos\phi$

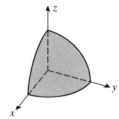

9. $V = \displaystyle\int_0^{2\pi} \int_0^3 \int_{r^2}^9 r\, dz\, dr\, d\theta = \int_0^{2\pi} \int_0^3 r(9-r^2)dr\, d\theta = \int_0^{2\pi} \frac{81}{4}d\theta = 81\pi/2$

11. $r^2 + z^2 = 20$ intersects $z = r^2$ in a circle of radius 2; the volume consists of two portions, one inside the cylinder $r = \sqrt{20}$ and one outside that cylinder:

$V = \displaystyle\int_0^{2\pi} \int_0^2 \int_{-\sqrt{20-r^2}}^{r^2} r\, dz\, dr\, d\theta + \int_0^{2\pi} \int_2^{\sqrt{20}} \int_{-\sqrt{20-r^2}}^{\sqrt{20-r^2}} r\, dz\, dr\, d\theta$

$= \displaystyle\int_0^{2\pi} \int_0^2 r\left(r^2 + \sqrt{20-r^2}\right)dr\, d\theta + \int_0^{2\pi} \int_2^{\sqrt{20}} 2r\sqrt{20-r^2}\, dr\, d\theta$

$= \dfrac{4}{3}(10\sqrt{5} - 13)\displaystyle\int_0^{2\pi} d\theta + \frac{128}{3}\int_0^{2\pi} d\theta = \frac{152}{3}\pi + \frac{80}{3}\pi\sqrt{5}$

13. $V = \displaystyle\int_0^{2\pi} \int_0^{\pi/3} \int_0^4 \rho^2 \sin\phi\, d\rho\, d\phi\, d\theta = \int_0^{2\pi} \int_0^{\pi/3} \frac{64}{3}\sin\phi\, d\phi\, d\theta = \frac{32}{3}\int_0^{2\pi} d\theta = 64\pi/3$

15. In spherical coordinates the sphere and the plane $z = a$ are $\rho = 2a$ and $\rho = a\sec\phi$, respectively. They intersect at $\phi = \pi/3$,

$$V = \int_0^{2\pi}\int_0^{\pi/3}\int_0^{a\sec\phi}\rho^2\sin\phi\,d\rho\,d\phi\,d\theta + \int_0^{2\pi}\int_{\pi/3}^{\pi/2}\int_0^{2a}\rho^2\sin\phi\,d\rho\,d\phi\,d\theta$$

$$= \int_0^{2\pi}\int_0^{\pi/3}\frac{1}{3}a^3\sec^3\phi\sin\phi\,d\phi\,d\theta + \int_0^{2\pi}\int_{\pi/3}^{\pi/2}\frac{8}{3}a^3\sin\phi\,d\phi\,d\theta$$

$$= \frac{1}{2}a^3\int_0^{2\pi}d\theta + \frac{4}{3}a^3\int_0^{2\pi}d\theta = 11\pi a^3/3$$

17. $$\int_0^{\pi/2}\int_0^a\int_0^{a^2-r^2}r^3\cos^2\theta\,dz\,dr\,d\theta = \int_0^{\pi/2}\int_0^a(a^2r^3 - r^5)\cos^2\theta\,dr\,d\theta$$

$$= \frac{1}{12}a^6\int_0^{\pi/2}\cos^2\theta\,d\theta = \pi a^6/48$$

19. $$\int_0^{\pi/2}\int_0^{\pi/4}\int_0^{\sqrt{8}}\rho^4\cos^2\phi\sin\phi\,d\rho\,d\phi\,d\theta = 32(2\sqrt{2}-1)\pi/15$$

21. (a) $$\int_{-2}^2\int_1^4\int_{\pi/6}^{\pi/3}\frac{r\tan^3\theta}{\sqrt{1+z^2}}\,d\theta\,dr\,dz = \left(\int_{-2}^2\frac{1}{\sqrt{1+z^2}}\,dz\right)\left(\int_1^4 r\,dr\right)\left(\int_{\pi/6}^{\pi/3}\tan^3\theta\,d\theta\right)$$

$$= \left(-2\ln(\sqrt{5}-2)\right)\frac{15}{2}\left(\frac{4}{3}-\frac{1}{2}\ln 3\right) \approx 16.97774196$$

The region is a cylindrical wedge.

(b) To convert to rectangular coordinates observe that the rays $\theta = \pi/6, \theta = \pi/3$ correspond to the lines $y = x/\sqrt{3}, y = \sqrt{3}x$. Then $dx\,dy\,dz = r\,dr\,d\theta\,dz$ and $\tan\theta = y/x$, hence

$$\text{Integral} = \int_1^4\int_{x/\sqrt{3}}^{\sqrt{3}x}\int_{-2}^2\frac{(y/x)^3}{\sqrt{1+z^2}}\,dz\,dy\,dx, \quad \text{so } f(x,y,z) = \frac{y^3}{x^3\sqrt{1+z^2}}.$$

23. (a) $$V = 2\int_0^{2\pi}\int_0^a\int_0^{\sqrt{a^2-r^2}}r\,dz\,dr\,d\theta = 4\pi a^3/3$$

(b) $$V = \int_0^{2\pi}\int_0^\pi\int_0^a\rho^2\sin\phi\,d\rho\,d\phi\,d\theta = 4\pi a^3/3$$

25. $$M = \int_0^{2\pi}\int_0^3\int_r^3(3-z)r\,dz\,dr\,d\theta = \int_0^{2\pi}\int_0^3\frac{1}{2}r(3-r)^2dr\,d\theta = \frac{27}{8}\int_0^{2\pi}d\theta = 27\pi/4$$

27. $$M = \int_0^{2\pi}\int_0^\pi\int_0^a k\rho^3\sin\phi\,d\rho\,d\phi\,d\theta = \int_0^{2\pi}\int_0^\pi\frac{1}{4}ka^4\sin\phi\,d\phi\,d\theta = \frac{1}{2}ka^4\int_0^{2\pi}d\theta = \pi ka^4$$

29. $\bar{x} = \bar{y} = 0$ from the symmetry of the region,

$$V = \int_0^{2\pi}\int_0^1\int_{r^2}^{\sqrt{2-r^2}}r\,dz\,dr\,d\theta = \int_0^{2\pi}\int_0^1(r\sqrt{2-r^2}-r^3)dr\,d\theta = (8\sqrt{2}-7)\pi/6,$$

$$\bar{z} = \frac{1}{V}\int_0^{2\pi}\int_0^1\int_{r^2}^{\sqrt{2-r^2}}zr\,dz\,dr\,d\theta = \frac{6}{(8\sqrt{2}-7)\pi}(7\pi/12) = 7/(16\sqrt{2}-14);$$

centroid $\left(0,0,\dfrac{7}{16\sqrt{2}-14}\right)$

31. $\bar{x} = \bar{y} = \bar{z}$ from the symmetry of the region, $V = \pi a^3/6$,

$$\bar{z} = \frac{1}{V} \int_0^{\pi/2} \int_0^{\pi/2} \int_0^a \rho^3 \cos\phi \sin\phi \, d\rho \, d\phi \, d\theta = \frac{6}{\pi a^3}(\pi a^4/16) = 3a/8;$$

centroid $(3a/8, 3a/8, 3a/8)$

33. $\bar{y} = 0$ from the symmetry of the region, $V = 2 \int_0^{\pi/2} \int_0^{2\cos\theta} \int_0^{r^2} r \, dz \, dr \, d\theta = 3\pi/2$,

$$\bar{x} = \frac{2}{V} \int_0^{\pi/2} \int_0^{2\cos\theta} \int_0^{r^2} r^2 \cos\theta \, dz \, dr \, d\theta = \frac{4}{3\pi}(\pi) = 4/3,$$

$$\bar{z} = \frac{2}{V} \int_0^{\pi/2} \int_0^{2\cos\theta} \int_0^{r^2} rz \, dz \, dr \, d\theta = \frac{4}{3\pi}(5\pi/6) = 10/9; \text{ centroid } (4/3, 0, 10/9)$$

35. $V = \int_0^{\pi/2} \int_{\pi/6}^{\pi/3} \int_0^2 \rho^2 \sin\phi \, d\rho \, d\phi \, d\theta = \int_0^{\pi/2} \int_{\pi/6}^{\pi/3} \frac{8}{3} \sin\phi \, d\phi \, d\theta = \frac{4}{3}(\sqrt{3} - 1) \int_0^{\pi/2} d\theta$

$\qquad = 2(\sqrt{3} - 1)\pi/3$

37. $\bar{x} = \bar{y} = 0$ from the symmetry of density and region,

$$M = \int_0^{2\pi} \int_0^1 \int_0^{1-r^2} (r^2 + z^2) r \, dz \, dr \, d\theta = \pi/4,$$

$$\bar{z} = \frac{1}{M} \int_0^{2\pi} \int_0^1 \int_0^{1-r^2} z(r^2+z^2) r \, dz \, dr \, d\theta = (4/\pi)(11\pi/120) = 11/30; \text{ center of gravity } (0, 0, 11/30)$$

39. $\bar{x} = \bar{y} = 0$ from the symmetry of density and region,

$$M = \int_0^{2\pi} \int_0^{\pi/2} \int_0^a k\rho^3 \sin\phi \, d\rho \, d\phi \, d\theta = \pi k a^4/2,$$

$$\bar{z} = \frac{1}{M} \int_0^{2\pi} \int_0^{\pi/2} \int_0^a k\rho^4 \sin\phi \cos\phi \, d\rho \, d\phi \, d\theta = \frac{2}{\pi k a^4}(\pi k a^5/5) = 2a/5; \text{ center of gravity } (0, 0, 2a/5)$$

41. $M = \int_0^{2\pi} \int_0^{\pi} \int_0^R \delta_0 e^{-(\rho/R)^3} \rho^2 \sin\phi \, d\rho \, d\phi \, d\theta = \int_0^{2\pi} \int_0^{\pi} \frac{1}{3}(1 - e^{-1})R^3 \delta_0 \sin\phi \, d\phi \, d\theta$

$\qquad = \frac{4}{3}\pi(1 - e^{-1})\delta_0 R^3$

43. $I_z = \int_0^{2\pi} \int_0^a \int_0^h r^2 \delta \, r \, dz \, dr \, d\theta = \delta \int_0^{2\pi} \int_0^a \int_0^h r^3 \, dz \, dr \, d\theta = \frac{1}{2}\delta\pi a^4 h$

45. $I_z = \int_0^{2\pi} \int_{a_1}^{a_2} \int_0^h r^2 \delta \, r \, dz \, dr \, d\theta = \delta \int_0^{2\pi} \int_{a_1}^{a_2} \int_0^h r^3 \, dz \, dr \, d\theta = \frac{1}{2}\delta\pi h(a_2^4 - a_1^4)$

EXERCISE SET 15.8

1. $\dfrac{\partial(x, y)}{\partial(u, v)} = \begin{vmatrix} 1 & 4 \\ 3 & -5 \end{vmatrix} = -17$

3. $\dfrac{\partial(x, y)}{\partial(u, v)} = \begin{vmatrix} \cos u & -\sin v \\ \sin u & \cos v \end{vmatrix} = \cos u \cos v + \sin u \sin v = \cos(u - v)$

5. $x = \dfrac{2}{9}u + \dfrac{5}{9}v,\ y = -\dfrac{1}{9}u + \dfrac{2}{9}v;\ \dfrac{\partial(x,y)}{\partial(u,v)} = \begin{vmatrix} 2/9 & 5/9 \\ -1/9 & 2/9 \end{vmatrix} = \dfrac{1}{9}$

7. $x = \sqrt{u+v}/\sqrt{2},\ y = \sqrt{v-u}/\sqrt{2};\ \dfrac{\partial(x,y)}{\partial(u,v)} = \begin{vmatrix} \dfrac{1}{2\sqrt{2}\sqrt{u+v}} & \dfrac{1}{2\sqrt{2}\sqrt{u+v}} \\ -\dfrac{1}{2\sqrt{2}\sqrt{v-u}} & \dfrac{1}{2\sqrt{2}\sqrt{v-u}} \end{vmatrix} = \dfrac{1}{4\sqrt{v^2 - u^2}}$

9. $\dfrac{\partial(x,y,z)}{\partial(u,v,w)} = \begin{vmatrix} 3 & 1 & 0 \\ 1 & 0 & -2 \\ 0 & 1 & 1 \end{vmatrix} = 5$

11. $y = v, x = u/y = u/v, z = w - x = w - u/v;\ \dfrac{\partial(x,y,z)}{\partial(u,v,w)} = \begin{vmatrix} 1/v & -u/v^2 & 0 \\ 0 & 1 & 0 \\ -1/v & u/v^2 & 1 \end{vmatrix} = 1/v$

13.

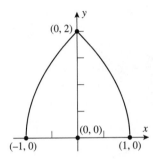

15.

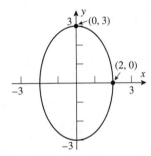

17. $x = \dfrac{1}{5}u + \dfrac{2}{5}v,\ y = -\dfrac{2}{5}u + \dfrac{1}{5}v,\ \dfrac{\partial(x,y)}{\partial(u,v)} = \dfrac{1}{5};\ \dfrac{1}{5}\iint\limits_S \dfrac{u}{v}\,dA_{uv} = \dfrac{1}{5}\int_1^3 \int_1^4 \dfrac{u}{v}\,du\,dv = \dfrac{3}{2}\ln 3$

19. $x = u + v,\ y = u - v,\ \dfrac{\partial(x,y)}{\partial(u,v)} = -2;$ the boundary curves of the region S in the uv-plane are

$v = 0, v = u,$ and $u = 1$ so $2\iint\limits_S \sin u \cos v\, dA_{uv} = 2\int_0^1 \int_0^u \sin u \cos v\, dv\, du = 1 - \dfrac{1}{2}\sin 2$

21. $x = 3u, y = 4v,\ \dfrac{\partial(x,y)}{\partial(u,v)} = 12;\ S$ is the region in the uv-plane enclosed by the circle $u^2 + v^2 = 1$.

Use polar coordinates to obtain $\iint\limits_S 12\sqrt{u^2 + v^2}(12)\,dA_{uv} = 144\int_0^{2\pi} \int_0^1 r^2 dr\, d\theta = 96\pi$

23. Let S be the region in the uv-plane bounded by $u^2 + v^2 = 1$, so $u = 2x, v = 3y$,

$x = u/2, y = v/3,\ \dfrac{\partial(x,y)}{\partial(u,v)} = \begin{vmatrix} 1/2 & 0 \\ 0 & 1/3 \end{vmatrix} = 1/6$, use polar coordinates to get

$\dfrac{1}{6}\iint\limits_S \sin(u^2 + v^2)du\, dv = \dfrac{1}{6}\int_0^{\pi/2} \int_0^1 r \sin r^2\, dr\, d\theta = \dfrac{\pi}{24}(-\cos r^2)\Big]_0^1 = \dfrac{\pi}{24}(1 - \cos 1)$

25. $x = u/3, y = v/2, z = w, \dfrac{\partial(x,y,z)}{\partial(u,v,w)} = 1/6$; S is the region in uvw-space enclosed by the sphere $u^2 + v^2 + w^2 = 36$ so

$$\iiint\limits_S \frac{u^2}{9}\frac{1}{6}\, dV_{uvw} = \frac{1}{54}\int_0^{2\pi}\int_0^{\pi}\int_0^6 (\rho\sin\phi\cos\theta)^2 \rho^2 \sin\phi\, d\rho\, d\phi\, d\theta$$

$$= \frac{1}{54}\int_0^{2\pi}\int_0^{\pi}\int_0^6 \rho^4 \sin^3\phi \cos^2\theta\, d\rho\, d\phi\, d\theta = \frac{192}{5}\pi$$

27. $u = \theta = \cot^{-1}(x/y), v = r = \sqrt{x^2 + y^2}$ $\qquad$ **29.** $u = \dfrac{3}{7}x - \dfrac{2}{7}y, v = -\dfrac{1}{7}x + \dfrac{3}{7}y$

31. Let $u = y - 4x, v = y + 4x$, then $x = \dfrac{1}{8}(v - u), y = \dfrac{1}{2}(v + u)$ so $\dfrac{\partial(x,y)}{\partial(u,v)} = -\dfrac{1}{8}$;

$$\frac{1}{8}\iint\limits_S \frac{u}{v}\, dA_{uv} = \frac{1}{8}\int_2^5\int_0^2 \frac{u}{v}\, du\, dv = \frac{1}{4}\ln\frac{5}{2}$$

33. Let $u = x - y, v = x + y$, then $x = \dfrac{1}{2}(v + u), y = \dfrac{1}{2}(v - u)$ so $\dfrac{\partial(x,y)}{\partial(u,v)} = \dfrac{1}{2}$; the boundary curves of the region S in the uv-plane are $u = 0, v = u$, and $v = \pi/4$; thus

$$\frac{1}{2}\iint\limits_S \frac{\sin u}{\cos v}\, dA_{uv} = \frac{1}{2}\int_0^{\pi/4}\int_0^v \frac{\sin u}{\cos v}\, du\, dv = \frac{1}{2}[\ln(\sqrt{2} + 1) - \pi/4]$$

35. Let $u = y/x, v = x/y^2$, then $x = 1/(u^2 v), y = 1/(uv)$ so $\dfrac{\partial(x,y)}{\partial(u,v)} = \dfrac{1}{u^4 v^3}$;

$$\iint\limits_S \frac{1}{u^4 v^3}\, dA_{uv} = \int_1^4\int_1^2 \frac{1}{u^4 v^3}\, du\, dv = 35/256$$

37. $x = u, y = w/u, z = v + w/u, \dfrac{\partial(x,y,z)}{\partial(u,v,w)} = -\dfrac{1}{u}$;

$$\iiint\limits_S \frac{v^2 w}{u}\, dV_{uvw} = \int_2^4\int_0^1\int_1^3 \frac{v^2 w}{u}\, du\, dv\, dw = 2\ln 3$$

39. $\dfrac{\partial(x,y,z)}{\partial(\rho,\phi,\theta)} = \begin{vmatrix} \sin^3\phi\cos^3\theta & 3\rho\sin^2\phi\cos\phi\cos^3\theta & -3\rho\sin^3\phi\cos^2\theta\sin\theta \\ \sin^3\phi\sin^3\theta & 3\rho\sin^2\phi\cos\phi\sin^3\theta & 3\rho\sin^3\phi\sin^2\theta\cos\theta \\ \cos^3\phi & -3\rho\cos^2\phi\sin\phi & 0 \end{vmatrix}$

$$= 9\rho^2\cos^2\theta\sin^2\theta\cos^2\phi\sin^5\phi,$$

$$V = 9\int_0^{2\pi}\int_0^{\pi}\int_0^a \rho^2\cos^2\theta\sin^2\theta\cos^2\phi\sin^5\phi\, d\rho\, d\phi\, d\theta = \frac{4}{35}\pi a^3$$

41. **(a)** $\dfrac{\partial(x,y)}{\partial(u,v)} = \begin{vmatrix} 1 - v & -u \\ v & u \end{vmatrix} = u$; $\quad u = x + y, v = \dfrac{y}{x + y}$,

$$\frac{\partial(u,v)}{\partial(x,y)} = \begin{vmatrix} 1 & 1 \\ -y/(x+y)^2 & x/(x+y)^2 \end{vmatrix} = \frac{x}{(x+y)^2} + \frac{y}{(x+y)^2} = \frac{1}{x+y} = \frac{1}{u};$$

$$\frac{\partial(u,v)}{\partial(x,y)}\frac{\partial(x,y)}{\partial(u,v)} = 1$$

(b) $\dfrac{\partial(x,y)}{\partial(u,v)} = \begin{vmatrix} v & u \\ 0 & 2v \end{vmatrix} = 2v^2; \quad u = x/\sqrt{y}, v = \sqrt{y},$

$$\frac{\partial(u,v)}{\partial(x,y)} = \begin{vmatrix} 1/\sqrt{y} & -x/(2y^{3/2}) \\ 0 & 1/(2\sqrt{y}) \end{vmatrix} = \frac{1}{2y} = \frac{1}{2v^2}; \frac{\partial(u,v)}{\partial(x,y)}\frac{\partial(x,y)}{\partial(u,v)} = 1$$

(c) $\dfrac{\partial(x,y)}{\partial(u,v)} = \begin{vmatrix} u & v \\ u & -v \end{vmatrix} = -2uv; \quad u = \sqrt{x+y}, v = \sqrt{x-y},$

$$\frac{\partial(u,v)}{\partial(x,y)} = \begin{vmatrix} 1/(2\sqrt{x+y}) & 1/(2\sqrt{x+y}) \\ 1/(2\sqrt{x-y}) & -1/(2\sqrt{x-y}) \end{vmatrix} = -\frac{1}{2\sqrt{x^2-y^2}} = -\frac{1}{2uv}; \frac{\partial(u,v)}{\partial(x,y)}\frac{\partial(x,y)}{\partial(u,v)} = 1$$

43. $\dfrac{\partial(u,v)}{\partial(x,y)} = 8xy$ so $\dfrac{\partial(x,y)}{\partial(u,v)} = \dfrac{1}{8xy}; \ xy\left|\dfrac{\partial(x,y)}{\partial(u,v)}\right| = xy\left(\dfrac{1}{8xy}\right) = \dfrac{1}{8}$ so

$$\frac{1}{8}\iint\limits_{S} dA_{uv} = \frac{1}{8}\int_9^{16}\int_1^4 du\, dv = 21/8$$

45. Set $u = x+y+2z, v = x-2y+z, w = 4x+y+z$, then $\dfrac{\partial(u,v,w)}{\partial(x,y,z)} = \begin{vmatrix} 1 & 1 & 2 \\ 1 & -2 & 1 \\ 4 & 1 & 1 \end{vmatrix} = 18$, and

$$V = \iiint\limits_{R} dx\, dy\, dz = \int_{-6}^{6}\int_{-2}^{2}\int_{-3}^{3} \frac{\partial(x,y,z)}{\partial(u,v,w)} du\, dv\, dw = 6(4)(12)\frac{1}{18} = 16$$

47. **(a)** $\dfrac{\partial(x,y,z)}{\partial(r,\theta,z)} = \begin{vmatrix} \cos\theta & -r\sin\theta & 0 \\ \sin\theta & r\cos\theta & 0 \\ 0 & 0 & 1 \end{vmatrix} = r, \ \left|\dfrac{\partial(x,y,z)}{\partial(r,\theta,z)}\right| = r$

(b) $\dfrac{\partial(x,y,z)}{\partial(\rho,\phi,\theta)} = \begin{vmatrix} \sin\phi\cos\theta & \rho\cos\phi\cos\theta & -\rho\sin\phi\sin\theta \\ \sin\phi\sin\theta & \rho\cos\phi\sin\theta & \rho\sin\phi\cos\theta \\ \cos\phi & -\rho\sin\phi & 0 \end{vmatrix} = \rho^2\sin\phi; \ \left|\dfrac{\partial(x,y,z)}{\partial(\rho,\phi,\theta)}\right| = \rho^2\sin\phi$

REVIEW EXERCISES, CHAPTER 15

3. **(a)** $\displaystyle\iint\limits_{R} dA$ **(b)** $\displaystyle\iiint\limits_{G} dV$ **(c)** $\displaystyle\iint\limits_{R}\sqrt{1+\left(\frac{\partial z}{\partial x}\right)^2 + \left(\frac{\partial z}{\partial y}\right)^2}\, dA$

7. $\displaystyle\int_0^1\int_{1-\sqrt{1-y^2}}^{1+\sqrt{1-y^2}} f(x,y)\, dx\, dy$

9. **(a)** $(1,2) = (b,d), (2,1) = (a,c)$, so $a = 2, b = 1, c = 1, d = 2$

(b) $\displaystyle\iint\limits_{R} dA = \int_0^1\int_0^1 \frac{\partial(x,y)}{\partial(u,v)} du\, dv = \int_0^1\int_0^1 3\, du\, dv = 3$

11. $\displaystyle\int_{1/2}^{1} 2x\cos(\pi x^2)\,dx = \frac{1}{\pi}\sin(\pi x^2)\Big]_{1/2}^{1} = -1/(\sqrt{2}\pi)$

13. $\displaystyle\int_{0}^{1}\int_{2y}^{2} e^x e^y\,dx\,dy$

15.

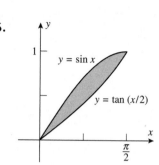

17. $\displaystyle 2\int_{0}^{8}\int_{0}^{y^{1/3}} x^2\sin y^2\,dx\,dy = \frac{2}{3}\int_{0}^{8} y\sin y^2\,dy = -\frac{1}{3}\cos y^2\Big]_{0}^{8} = \frac{1}{3}(1-\cos 64) \approx 0.20271$

19. $\sin 2\theta = 2\sin\theta\cos\theta = \dfrac{2xy}{x^2+y^2}$, and $r = 2a\sin\theta$ is the circle $x^2 + (y-a)^2 = a^2$, so

$$\int_{0}^{a}\int_{a-\sqrt{a^2-x^2}}^{a+\sqrt{a^2-x^2}} \frac{2xy}{x^2+y^2}\,dy\,dx = \int_{0}^{a} x\left[\ln\left(a+\sqrt{a^2-x^2}\right) - \ln\left(a-\sqrt{a^2-x^2}\right)\right]dx = a^2$$

21. $\displaystyle\int_{0}^{2}\int_{(y/2)^{1/3}}^{2-y/2} dx\,dy = \int_{0}^{2}\left(2 - \frac{y}{2} - \left(\frac{y}{2}\right)^{1/3}\right)dy = \left(2y - \frac{y^2}{4} - \frac{3}{2}\left(\frac{y}{2}\right)^{4/3}\right)\Big]_{0}^{2} = \frac{3}{2}$

23. $\displaystyle\int_{0}^{2\pi}\int_{0}^{2}\int_{r^4}^{16} r^2\cos^2\theta\,r\,dz\,dr\,d\theta = \int_{0}^{2\pi}\cos^2\theta\,d\theta\int_{0}^{2} r^3(16-r^4)\,dr = 32\pi$

25. **(a)** $\displaystyle\int_{0}^{2\pi}\int_{0}^{\pi/3}\int_{0}^{a}(\rho^2\sin^2\phi)\rho^2\sin\phi\,d\rho\,d\phi\,d\theta = \int_{0}^{2\pi}\int_{0}^{\pi/3}\int_{0}^{a}\rho^4\sin^3\phi\,d\rho\,d\phi\,d\theta$

(b) $\displaystyle\int_{0}^{2\pi}\int_{0}^{\sqrt{3}a/2}\int_{r/\sqrt{3}}^{\sqrt{a^2-r^2}} r^2\,dz\,r\,dr\,d\theta = \int_{0}^{2\pi}\int_{0}^{\sqrt{3}a/2}\int_{r/\sqrt{3}}^{\sqrt{a^2-r^2}} r^3\,dz\,dr\,d\theta$

(c) $\displaystyle\int_{-\sqrt{3}a/2}^{\sqrt{3}a/2}\int_{-\sqrt{(3a^2/4)-x^2}}^{\sqrt{(3a^2/4)-x^2}}\int_{\sqrt{x^2+y^2}/\sqrt{3}}^{\sqrt{a^2-x^2-y^2}}(x^2+y^2)\,dz\,dy\,dx$

27. $\displaystyle V = \int_{0}^{2\pi}\int_{0}^{a/\sqrt{3}}\int_{\sqrt{3}r}^{a} r\,dz\,dr\,d\theta = 2\pi\int_{0}^{a/\sqrt{3}} r(a-\sqrt{3}r)\,dr = \frac{\pi a^3}{9}$

29. $\|\mathbf{r}_u \times \mathbf{r}_v\| = \sqrt{2u^2 + 2v^2 + 4}$,

$$S = \iint\limits_{u^2+v^2\le 4} \sqrt{2u^2+2v^2+4}\,dA = \int_{0}^{2\pi}\int_{0}^{2}\sqrt{2}\sqrt{r^2+2}\,r\,dr\,d\theta = \frac{8\pi}{3}(3\sqrt{3}-1)$$

31. $(\mathbf{r}_u \times \mathbf{r}_v)\Big]_{\substack{u=1\\v=2}} = \langle -2,-4,1\rangle$, tangent plane $2x+4y-z=5$

33. $\displaystyle A = \int_{-4}^{4}\int_{y^2/4}^{2+y^2/8} dx\,dy = \int_{-4}^{4}\left(2 - \frac{y^2}{8}\right)dy = \frac{32}{3}; \bar{y} = 0$ by symmetry;

$$\int_{-4}^{4} \int_{y^2/4}^{2+y^2/8} x \, dx \, dy = \int_{-4}^{4} \left(2 + \frac{1}{4}y^2 - \frac{3}{128}y^4\right) dy = \frac{256}{15}, \ \bar{x} = \frac{3}{32}\frac{256}{15} = \frac{8}{5}; \ \text{centroid} \left(\frac{8}{5}, 0\right)$$

35. $V = \dfrac{1}{3}\pi a^2 h, \bar{x} = \bar{y} = 0$ by symmetry,

$$\int_{0}^{2\pi} \int_{0}^{a} \int_{0}^{h - rh/a} rz \, dz \, dr \, d\theta = \pi \int_{0}^{a} rh^2 \left(1 - \frac{r}{a}\right)^2 dr = \pi a^2 h^2 /12, \ \text{centroid} \ (0, 0, h/4)$$

37. $V = \dfrac{4}{3}\pi a^3, \bar{d} = \dfrac{3}{4\pi a^3} \iiint\limits_{\rho \leq a} \rho dV = \dfrac{3}{4\pi a^3} \int_{0}^{\pi} \int_{0}^{2\pi} \int_{0}^{a} \rho^3 \sin\phi \, d\rho \, d\theta \, d\phi = \dfrac{3}{4\pi a^3} 2\pi(2)\dfrac{a^4}{4} = \dfrac{3}{4}a$

39. (a) Add u and w to get $x = \ln(u + w) - \ln 2$; subtract w from u to get $y = \dfrac{1}{2}u - \dfrac{1}{2}w$, substitute these values into $v = y + 2z$ to get $z = -\dfrac{1}{4}u + \dfrac{1}{2}v + \dfrac{1}{4}w$. Hence $x_u = \dfrac{1}{u+w}, x_v = 0, x_w = \dfrac{1}{u+w}; y_u = \dfrac{1}{2}, y_v = 0, y_z = -\dfrac{1}{2}; z_u = -\dfrac{1}{4}, z_v = \dfrac{1}{2}, z_w = \dfrac{1}{4}$, and thus $\dfrac{\partial(x, y, z)}{\partial(u, v, w)} = \dfrac{1}{2(u + w)}$

(b) $V = \iiint\limits_{G} = \int_{1}^{3} \int_{1}^{2} \int_{0}^{4} \dfrac{1}{2(u + w)} \, dw \, dv \, du$

$= (7\ln 7 - 5\ln 5 - 3\ln 3)/2 = \dfrac{1}{2}\ln\dfrac{823543}{84375} \approx 1.139172308$

Topics in Vector Calculus

EXERCISE SET 16.1

1. (a) III because the vector field is independent of y and the direction is that of the negative x-axis for negative x, and positive for positive

(b) IV, because the y-component is constant, and the x-component varies periodically with x

3. (a) true   **(b)** true   **(c)** true

5. 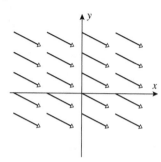   **7.**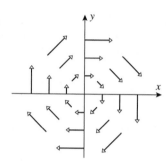

9.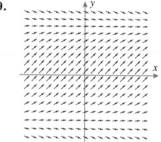

11. (a) $\nabla\phi = \phi_x \mathbf{i} + \phi_y \mathbf{j} = \dfrac{y}{1+x^2 y^2}\mathbf{i} + \dfrac{x}{1+x^2 y^2}\mathbf{j} = \mathbf{F}$, so $\mathbf{F}$ is conservative for all x, y

(b) $\nabla\phi = \phi_x \mathbf{i} + \phi_y \mathbf{j} = 2x\mathbf{i} - 6y\mathbf{j} + 8z\mathbf{k} = \mathbf{F}$ so $\mathbf{F}$ is conservative for all x, y

13. div $\mathbf{F} = 2x + y$, curl $\mathbf{F} = z\mathbf{i}$

15. div $\mathbf{F} = 0$, curl $\mathbf{F} = (40x^2 z^4 - 12xy^3)\mathbf{i} + (14y^3 z + 3y^4)\mathbf{j} - (16xz^5 + 21y^2 z^2)\mathbf{k}$

17. div $\mathbf{F} = \dfrac{2}{\sqrt{x^2 + y^2 + z^2}}$, curl $\mathbf{F} = \mathbf{0}$

19. $\nabla \cdot (\mathbf{F} \times \mathbf{G}) = \nabla \cdot (-(z + 4y^2)\mathbf{i} + (4xy + 2xz)\mathbf{j} + (2xy - x)\mathbf{k}) = 4x$

21. $\nabla \cdot (\nabla \times \mathbf{F}) = \nabla \cdot (-\sin(x - y)\mathbf{k}) = 0$

23. $\nabla \times (\nabla \times \mathbf{F}) = \nabla \times (xz\mathbf{i} - yz\mathbf{j} + y\mathbf{k}) = (1 + y)\mathbf{i} + x\mathbf{j}$

27. Let $\mathbf{F} = f\mathbf{i} + g\mathbf{j} + h\mathbf{k}$; div $(k\mathbf{F}) = k\dfrac{\partial f}{\partial x} + k\dfrac{\partial g}{\partial y} + k\dfrac{\partial h}{\partial z} = k$ div $\mathbf{F}$

29. Let $\mathbf{F} = f(x, y, z)\mathbf{i} + g(x, y, z)\mathbf{j} + h(x, y, z)\mathbf{k}$ and $\mathbf{G} = P(x, y, z)\mathbf{i} + Q(x, y, z)\mathbf{j} + R(x, y, z)\mathbf{k}$, then

$$\text{div } (\mathbf{F} + \mathbf{G}) = \left(\frac{\partial f}{\partial x} + \frac{\partial P}{\partial x}\right) + \left(\frac{\partial g}{\partial y} + \frac{\partial Q}{\partial y}\right) + \left(\frac{\partial h}{\partial z} + \frac{\partial R}{\partial z}\right)$$

$$= \left(\frac{\partial f}{\partial x} + \frac{\partial g}{\partial y} + \frac{\partial h}{\partial z}\right) + \left(\frac{\partial P}{\partial x} + \frac{\partial Q}{\partial y} + \frac{\partial R}{\partial z}\right) = \text{div } \mathbf{F} + \text{div } \mathbf{G}$$

31. Let $\mathbf{F} = f\mathbf{i} + g\mathbf{j} + h\mathbf{k}$;

$$\text{div } (\phi\mathbf{F}) = \left(\phi\frac{\partial f}{\partial x} + \frac{\partial \phi}{\partial x}f\right) + \left(\phi\frac{\partial g}{\partial y} + \frac{\partial \phi}{\partial y}g\right) + \left(\phi\frac{\partial h}{\partial z} + \frac{\partial \phi}{\partial z}h\right)$$

$$= \phi\left(\frac{\partial f}{\partial x} + \frac{\partial g}{\partial y} + \frac{\partial h}{\partial z}\right) + \left(\frac{\partial \phi}{\partial x}f + \frac{\partial \phi}{\partial y}g + \frac{\partial \phi}{\partial z}h\right)$$

$$= \phi \text{ div } \mathbf{F} + \nabla\phi \cdot \mathbf{F}$$

33. Let $\mathbf{F} = f\mathbf{i} + g\mathbf{j} + h\mathbf{k}$;

$$\text{div(curl } \mathbf{F}) = \frac{\partial}{\partial x}\left(\frac{\partial h}{\partial y} - \frac{\partial g}{\partial z}\right) + \frac{\partial}{\partial y}\left(\frac{\partial f}{\partial z} - \frac{\partial h}{\partial x}\right) + \frac{\partial}{\partial z}\left(\frac{\partial g}{\partial x} - \frac{\partial f}{\partial y}\right)$$

$$= \frac{\partial^2 h}{\partial x \partial y} - \frac{\partial^2 g}{\partial x \partial z} + \frac{\partial^2 f}{\partial y \partial z} - \frac{\partial^2 h}{\partial y \partial x} + \frac{\partial^2 g}{\partial z \partial x} - \frac{\partial^2 f}{\partial z \partial y} = 0,$$

assuming equality of mixed second partial derivatives

35. $\nabla \cdot (k\mathbf{F}) = k\nabla \cdot \mathbf{F}$, $\nabla \cdot (\mathbf{F} + \mathbf{G}) = \nabla \cdot \mathbf{F} + \nabla \cdot \mathbf{G}$, $\nabla \cdot (\phi\mathbf{F}) = \phi\nabla \cdot \mathbf{F} + \nabla\phi \cdot \mathbf{F}$, $\nabla \cdot (\nabla \times \mathbf{F}) = 0$

37. **(a)** curl $\mathbf{r} = 0\mathbf{i} + 0\mathbf{j} + 0\mathbf{k} = \mathbf{0}$

(b) $\nabla\|\mathbf{r}\| = \nabla\sqrt{x^2 + y^2 + z^2} = \dfrac{x}{\sqrt{x^2 + y^2 + z^2}}\mathbf{i} + \dfrac{y}{\sqrt{x^2 + y^2 + z^2}}\mathbf{j} + \dfrac{z}{\sqrt{x^2 + y^2 + z^2}}\mathbf{k} = \dfrac{\mathbf{r}}{\|\mathbf{r}\|}$

39. **(a)** $\nabla f(r) = f'(r)\dfrac{\partial r}{\partial x}\mathbf{i} + f'(r)\dfrac{\partial r}{\partial y}\mathbf{j} + f'(r)\dfrac{\partial r}{\partial z}\mathbf{k} = f'(r)\nabla r = \dfrac{f'(r)}{r}\mathbf{r}$

(b) $\text{div}[f(r)\mathbf{r}] = f(r)\text{div } \mathbf{r} + \nabla f(r) \cdot \mathbf{r} = 3f(r) + \dfrac{f'(r)}{r}\mathbf{r} \cdot \mathbf{r} = 3f(r) + rf'(r)$

41. $f(r) = 1/r^3$, $f'(r) = -3/r^4$, $\text{div}(\mathbf{r}/r^3) = 3(1/r^3) + r(-3/r^4) = 0$

43. **(a)** At the point (x, y) the slope of the line along which the vector $-y\mathbf{i} + x\mathbf{j}$ lies is $-x/y$; the slope of the tangent line to C at (x, y) is dy/dx, so $dy/dx = -x/y$.

(b) $y\,dy = -x\,dx$, $y^2/2 = -x^2/2 + K_1$, $x^2 + y^2 = K$

45. $dy/dx = 1/x, y = \ln x + K$

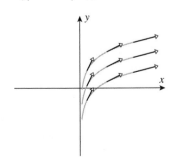

EXERCISE SET 16.2

1. **(a)** $\displaystyle\int_0^1 dy = 1$ because $s = y$ is arclength measured from $(0,0)$

   (b) 0, because $\sin xy = 0$ along C

3. Since $\mathbf{F}$ and $\mathbf{r}$ are parallel, $\mathbf{F} \cdot \mathbf{r} = \|\mathbf{F}\|\|\mathbf{r}\|$, and since $\mathbf{F}$ is constant,

$$\int_C \mathbf{F} \cdot d\mathbf{r} = \int_C d(\mathbf{F} \cdot \mathbf{r}) = \int_C d(\|\mathbf{F}\|\|\mathbf{r}\|) = \sqrt{2}\int_{-4}^4 \sqrt{2}\,dt = 16$$

5. By inspection the tangent vector in part (a) is given by $\mathbf{T} = \mathbf{j}$, so $\mathbf{F} \cdot \mathbf{T} = \mathbf{F} \cdot \mathbf{j} = \sin x$ on C. But $x = -\pi/2$ on C, thus $\sin x = -1$, $\mathbf{F} \cdot \mathbf{T} = -1$ and $\displaystyle\int_C \mathbf{F} \cdot d\mathbf{r} = \int_C (-1)ds$.

7. **(a)** $ds = \sqrt{\left(\dfrac{dx}{dt}\right)^2 + \left(\dfrac{dy}{dt}\right)^2}\,dt$, so $\displaystyle\int_0^1 (2t - 3t^2)\sqrt{4 + 36t^2}\,dt = -\dfrac{11}{108}\sqrt{10} - \dfrac{1}{36}\ln(\sqrt{10} - 3) - \dfrac{4}{27}$

   (b) $\displaystyle\int_0^1 (2t - 3t^2)2\,dt = 0$ **(c)** $\displaystyle\int_0^1 (2t - 3t^2)6t\,dt = -\dfrac{1}{2}$

9. **(a)** $C : x = t,\ y = t,\ 0 \le t \le 1$; $\displaystyle\int_0^1 6t\,dt = 3$

   (b) $C : x = t,\ y = t^2,\ 0 \le t \le 1$; $\displaystyle\int_0^1 (3t + 6t^2 - 2t^3)dt = 3$

   (c) $C : x = t,\ y = \sin(\pi t/2),\ 0 \le t \le 1$;
   $$\int_0^1 [3t + 2\sin(\pi t/2) + \pi t\cos(\pi t/2) - (\pi/2)\sin(\pi t/2)\cos(\pi t/2)]dt = 3$$

   (d) $C : x = t^3,\ y = t,\ 0 \le t \le 1$; $\displaystyle\int_0^1 (9t^5 + 8t^3 - t)dt = 3$

11. $\displaystyle\int_0^3 \dfrac{\sqrt{1+t}}{1+t}\,dt = \int_0^3 (1+t)^{-1/2}\,dt = 2$

13. $\displaystyle\int_0^1 3(t^2)(t^2)(2t^3/3)(1 + 2t^2)\,dt = 2\int_0^1 t^7(1 + 2t^2)\,dt = 13/20$

15. $\displaystyle\int_0^{\pi/4} (8\cos^2 t - 16\sin^2 t - 20\sin t\cos t)dt = 1 - \pi$

17. $C : x = (3-t)^2/3,\ y = 3 - t,\ 0 \le t \le 3$; $\displaystyle\int_0^3 \dfrac{1}{3}(3-t)^2 dt = 3$

19. $C : x = \cos t,\ y = \sin t,\ 0 \le t \le \pi/2$; $\displaystyle\int_0^{\pi/2} (-\sin t - \cos^2 t)dt = -1 - \pi/4$

21. $\displaystyle\int_0^1 (-3)e^{3t}dt = 1 - e^3$

23. **(a)** $\displaystyle\int_0^{\ln 2} \left(e^{3t} + e^{-3t}\right)\sqrt{e^{2t} + e^{-2t}}\,dt = \dfrac{63}{64}\sqrt{17} + \dfrac{1}{4}\ln(4 + \sqrt{17}) - \dfrac{1}{4}\tanh^{-1}\left(\dfrac{1}{17}\sqrt{17}\right)$

(b) $\displaystyle\int_0^{\pi/2} \left[e^t \sin t \cos t - (\sin t - t) \sin t + (1 + t^2) \right] dt = \frac{1}{24}\pi^3 + \frac{1}{5}e^{\pi/2} + \frac{1}{4}\pi + \frac{6}{5}$

25. **(a)** $C_1 : (0,0)$ to $(1,0); x = t, y = 0, 0 \le t \le 1$

$C_2 : (1,0)$ to $(0,1); x = 1 - t, y = t, 0 \le t \le 1$

$C_3 : (0,1)$ to $(0,0); x = 0, y = 1 - t, 0 \le t \le 1$

$\displaystyle\int_0^1 (0)dt + \int_0^1 (-1)dt + \int_0^1 (0)dt = -1$

(b) $C_1 : (0,0)$ to $(1,0); x = t, y = 0, 0 \le t \le 1$

$C_2 : (1,0)$ to $(1,1); x = 1, y = t, 0 \le t \le 1$

$C_3 : (1,1)$ to $(0,1); x = 1 - t, y = 1, 0 \le t \le 1$

$C_4 : (0,1)$ to $(0,0); x = 0, y = 1 - t, 0 \le t \le 1$

$\displaystyle\int_0^1 (0)dt + \int_0^1 (-1)dt + \int_0^1 (-1)dt + \int_0^1 (0)dt = -2$

27. $C_1 : x = t, y = z = 0, 0 \le t \le 1, \displaystyle\int_0^1 0\, dt = 0; \quad C_2 : x = 1, y = t, z = 0, 0 \le t \le 1, \displaystyle\int_0^1 (-t)\, dt = -\frac{1}{2}$

$C_3 : x = 1, y = 1, z = t, 0 \le t \le 1, \displaystyle\int_0^1 3\, dt = 3; \quad \displaystyle\int_C x^2 z\, dx - yx^2\, dy + 3\, dz = 0 - \frac{1}{2} + 3 = \frac{5}{2}$

29. $\displaystyle\int_0^\pi (0)dt = 0$

31. $\displaystyle\int_0^1 e^{-t}dt = 1 - e^{-1}$

33. Represent the circular arc by $x = 3\cos t, y = 3\sin t, 0 \le t \le \pi/2$.

$\displaystyle\int_C x\sqrt{y}\,ds = 9\sqrt{3}\int_0^{\pi/2}\sqrt{\sin t}\cos t\, dt = 6\sqrt{3}$

35. $\displaystyle\int_C \frac{kx}{1+y^2}ds = 15k\int_0^{\pi/2}\frac{\cos t}{1 + 9\sin^2 t}\, dt = 5k\tan^{-1} 3$

37. $C : x = t^2, y = t, 0 \le t \le 1; W = \displaystyle\int_0^1 3t^4 dt = 3/5$

41. $C : x = 4\cos t, y = 4\sin t, 0 \le t \le \pi/2$

$\displaystyle\int_0^{\pi/2}\left(-\frac{1}{4}\sin t + \cos t\right) dt = 3/4$

43. Represent the parabola by $x = t, y = t^2, 0 \le t \le 2$.

$\displaystyle\int_C 3x\, ds = \int_0^2 3t\sqrt{1 + 4t^2}\, dt = (17\sqrt{17} - 1)/4$

45. **(a)** $2\pi r h = 2\pi(1)2 = 4\pi$

(b) $S = \displaystyle\int_C z(t)\, dt$

(c) $C : x = \cos t, y = \sin t, 0 \le t \le 2\pi; S = \displaystyle\int_0^{2\pi}(2 + (1/2)\sin 3t)\, dt = 4\pi$

47. $W = \displaystyle\int_C \mathbf{F} \cdot d\mathbf{r} = \int_0^1 (\lambda t^2(1-t), t - \lambda t(1-t)) \cdot (1, \lambda - 2\lambda t)\, dt = -\lambda/12$, $W = 1$ when $\lambda = -12$

49. **(a)** From (8), $\Delta s_k = \displaystyle\int_{t_{k-1}}^{t_k} \|\mathbf{r}'(t)\|\, dt$, thus $m\Delta t_k \le \Delta s_k \le M\Delta t_k$ for all k. Obviously

$\Delta s_k \le M(\max\Delta t_k)$, and since the right side of this inequality is independent of k, it follows that $\max\Delta s_k \le M(\max\Delta t_k)$. Similarly $m(\max\Delta t_k) \le \max\Delta s_k$.

(b) This follows from $\max\Delta t_k \le \dfrac{1}{m}\max\Delta s_k$ and $\max\Delta s_k \le M\max\Delta t_k$.

EXERCISE SET 16.3

1. $\partial x/\partial y = 0 = \partial y/\partial x$, conservative so $\partial\phi/\partial x = x$ and $\partial\phi/\partial y = y$, $\phi = x^2/2 + k(y)$, $k'(y) = y$, $k(y) = y^2/2 + K$, $\phi = x^2/2 + y^2/2 + K$

3. $\partial(x^2 y)/\partial y = x^2$ and $\partial(5xy^2)/\partial x = 5y^2$, not conservative

5. $\partial(\cos y + y\cos x)/\partial y = -\sin y + \cos x = \partial(\sin x - x\sin y)/\partial x$, conservative so $\partial\phi/\partial x = \cos y + y\cos x$ and $\partial\phi/\partial y = \sin x - x\sin y$, $\phi = x\cos y + y\sin x + k(y)$, $-x\sin y + \sin x + k'(y) = \sin x - x\sin y$, $k'(y) = 0$, $k(y) = K$, $\phi = x\cos y + y\sin x + K$

7. **(a)** $\partial(y^2)/\partial y = 2y = \partial(2xy)/\partial x$, independent of path

(b) $C : x = -1 + 2t$, $y = 2 + t$, $0 \le t \le 1$; $\displaystyle\int_0^1 (4 + 14t + 6t^2)\, dt = 13$

(c) $\partial\phi/\partial x = y^2$ and $\partial\phi/\partial y = 2xy$, $\phi = xy^2 + k(y)$, $2xy + k'(y) = 2xy$, $k'(y) = 0$, $k(y) = K$, $\phi = xy^2 + K$. Let $K = 0$ to get $\phi(1,3) - \phi(-1,2) = 9 - (-4) = 13$

9. $\partial(3y)/\partial y = 3 = \partial(3x)/\partial x$, $\phi = 3xy$, $\phi(4,0) - \phi(1,2) = -6$

11. $\partial(2xe^y)/\partial y = 2xe^y = \partial(x^2 e^y)/\partial x$, $\phi = x^2 e^y$, $\phi(3,2) - \phi(0,0) = 9e^2$

13. $\partial(2xy^3)/\partial y = 6xy^2 = \partial(3x^2 y^2)/\partial x$, $\phi = x^2 y^3$, $\phi(-1,0) - \phi(2,-2) = 32$

15. $\phi = x^2 y^2/2$, $W = \phi(0,0) - \phi(1,1) = -1/2$

17. $\phi = e^{xy}$, $W = \phi(2,0) - \phi(-1,1) = 1 - e^{-1}$

19. $\partial(e^y + ye^x)/\partial y = e^y + e^x = \partial(xe^y + e^x)/\partial x$ so $\mathbf{F}$ is conservative, $\phi(x,y) = xe^y + ye^x$ so

$\displaystyle\int_C \mathbf{F} \cdot d\mathbf{r} = \phi(0, \ln 2) - \phi(1,0) = \ln 2 - 1$

21. $\mathbf{F} \cdot d\mathbf{r} = [(e^y + ye^x)\mathbf{i} + (xe^y + e^x)\mathbf{j}] \cdot [(\pi/2)\cos(\pi t/2)\mathbf{i} + (1/t)\mathbf{j}]dt$

$= \left(\dfrac{\pi}{2}\cos(\pi t/2)(e^y + ye^x) + (xe^y + e^x)/t\right) dt,$

so $\displaystyle\int_C \mathbf{F} \cdot d\mathbf{r} = \int_1^2 \left(\dfrac{\pi}{2}\cos(\pi t/2)\left(t + (\ln t)e^{\sin(\pi t/2)}\right) + \left(\sin(\pi t/2) + \dfrac{1}{t}e^{\sin(\pi t/2)}\right)\right) dt = \ln 2 - 1$

23. No; a closed loop can be found whose tangent everywhere makes an angle $< \pi$ with the vector field, so the line integral $\displaystyle\int_C \mathbf{F} \cdot d\mathbf{r} > 0$, and by Theorem 16.3.2 the vector field is not conservative.

25. Let $\mathbf{r}(t)$ be a parametrization of the circle C. Then by Theorem 16.3.2(b),

$\displaystyle\int_C \mathbf{F}\,d\mathbf{r} = \int_C \mathbf{F}\cdot\mathbf{r}'(t)\,dt = 0$. Let $h(t) = \mathbf{F}(x,y)\cdot\mathbf{r}'(t)$. Then h is continuous. We must find two points at which $h = 0$. If $h(t) = 0$ everywhere on the circle, then we are done; otherwise there are points at which h is nonzero, say $h(t_1) > 0$. Then there is a small interval around t_1 on which the integral of h is positive.

(Let $\epsilon = h(t_1)/2$. Since $h(t)$ is continuous there exists $\delta > 0$ such that for $|t - t_1| < \delta$, $h(t) > \epsilon/2$.

Then $\displaystyle\int_{t_1-\delta}^{t_1+\delta} h(t)\,dt \geq (2\delta)\epsilon/2 > 0$.)

Since $\displaystyle\int_C h = 0$, there are points on the circle where $h < 0$, say $h(t_2) < 0$. Now consider the parametrization $h(\theta)$, $0 \leq \theta \leq 2\pi$. Let $\theta_1 < \theta_2$ correspond to the points above where $h > 0$ and $h < 0$. Then by the Intermediate Value Theorem on $[\theta_1, \theta_2]$ there must be a point where $h = 0$, say $h(\theta_3) = 0, \theta_1 < \theta_3 < \theta2$.

To find a second point where $h = 0$, assume that h is a periodic function with period 2π (if need be, extend the definition of h). Then $h(t_2 - 2\pi) = h(t_2) < 0$. Apply the Intermediate Value Theorem on $[t_2 - 2\pi, t_1]$ to find an additional point θ_4 at which $h = 0$. The reader should prove that θ_3 and θ_4 do indeed correspond to distinct points on the circle.

27. If $\mathbf{F}$ is conservative, then $\mathbf{F} = \nabla\phi = \dfrac{\partial\phi}{\partial x}\mathbf{i} + \dfrac{\partial\phi}{\partial y}\mathbf{j} + \dfrac{\partial\phi}{\partial z}\mathbf{k}$ and hence $f = \dfrac{\partial\phi}{\partial x}, g = \dfrac{\partial\phi}{\partial y}$, and $h = \dfrac{\partial\phi}{\partial z}$.

Thus $\dfrac{\partial f}{\partial y} = \dfrac{\partial^2\phi}{\partial y\partial x}$ and $\dfrac{\partial g}{\partial x} = \dfrac{\partial^2\phi}{\partial x\partial y}, \dfrac{\partial f}{\partial z} = \dfrac{\partial^2\phi}{\partial z\partial x}$ and $\dfrac{\partial h}{\partial x} = \dfrac{\partial^2\phi}{\partial x\partial z}, \dfrac{\partial g}{\partial z} = \dfrac{\partial^2\phi}{\partial z\partial y}$ and $\dfrac{\partial h}{\partial y} = \dfrac{\partial^2\phi}{\partial y\partial z}$.

The result follows from the equality of mixed second partial derivatives.

29. $\dfrac{\partial}{\partial y}(h(x)[x\sin y + y\cos y]) = h(x)[x\cos y - y\sin y + \cos y]$

$\dfrac{\partial}{\partial x}(h(x)[x\cos y - y\sin y]) = h(x)\cos y + h'(x)[x\cos y - y\sin y]$,

equate these two partial derivatives to get $(x\cos y - y\sin y)(h'(x) - h(x)) = 0$ which holds for all x and y if $h'(x) = h(x)$, $h(x) = Ce^x$ where C is an arbitrary constant.

31. (a) See Exercise 30, $c = 1$; $W = \displaystyle\int_P^Q \mathbf{F}\cdot d\mathbf{r} = \phi(3,2,1) - \phi(1,1,2) = -\dfrac{1}{\sqrt{14}} + \dfrac{1}{\sqrt{6}}$

(b) C begins at $P(1,1,2)$ and ends at $Q(3,2,1)$ so the answer is again $W = -\dfrac{1}{\sqrt{14}} + \dfrac{1}{\sqrt{6}}$.

(c) The circle is not specified, but cannot pass through $(0,0,0)$, so Φ is continuous and differentiable on the circle. Start at any point P on the circle and return to P, so the work is $\Phi(P) - \Phi(P) = 0$.
C begins at, say, $(3,0)$ and ends at the same point so $W = 0$.

33. If C is composed of smooth curves $C_1, C_2, \ldots, C_n$ and curve C_i extends from (x_{i-1}, y_{i-1}) to (x_i, y_i)

then $\displaystyle\int_C \mathbf{F}\cdot d\mathbf{r} = \sum_{i=1}^n \int_{C_i} \mathbf{F}\cdot d\mathbf{r} = \sum_{i=1}^n [\phi(x_i, y_i) - \phi(x_{i-1}, y_{i-1})] = \phi(x_n, y_n) - \phi(x_0, y_0)$

where (x_0, y_0) and (x_n, y_n) are the endpoints of C.

35. Let C_1 be an arbitrary piecewise smooth curve from (a, b) to a point (x, y_1) in the disk, and C_2 the vertical line segment from (x, y_1) to (x, y). Then

$$\phi(x,y) = \int_{C_1} \mathbf{F} \cdot d\mathbf{r} + \int_{C_2} \mathbf{F} \cdot d\mathbf{r} = \int_{(a,b)}^{(x,y_1)} \mathbf{F} \cdot d\mathbf{r} + \int_{C_2} \mathbf{F} \cdot d\mathbf{r}.$$

The first term does not depend on y;

hence $\dfrac{\partial \phi}{\partial y} = \dfrac{\partial}{\partial y} \displaystyle\int_{C_2} \mathbf{F} \cdot d\mathbf{r} = \dfrac{\partial}{\partial y} \displaystyle\int_{C_2} f(x,y)dx + g(x,y)dy.$

However, the line integral with respect to x is zero along C_2, so $\dfrac{\partial \phi}{\partial y} = \dfrac{\partial}{\partial y} \displaystyle\int_{C_2} g(x,y)\ dy.$

Express C_2 as $x = x, y = t$ where t varies from y_1 to y, then $\dfrac{\partial \phi}{\partial y} = \dfrac{\partial}{\partial y} \displaystyle\int_{y_1}^{y} g(x,t)\ dt = g(x,y).$

EXERCISE SET 16.4

1. $\displaystyle\iint\limits_{R} (2x - 2y)dA = \int_0^1 \int_0^1 (2x - 2y)dy\ dx = 0$; for the line integral, on $x = 0, y^2\ dx = 0, x^2\ dy = 0$;

on $y = 0, y^2\ dx = x^2\ dy = 0$; on $x = 1, y^2\ dx + x^2\ dy = dy$; and on $y = 1, y^2\ dx + x^2\ dy = dx$,

hence $\displaystyle\oint\limits_{C} y^2\ dx + x^2\ dy = \int_0^1 dy + \int_1^0 dx = 1 - 1 = 0$

3. $\displaystyle\int_{-2}^{4} \int_1^2 (2y - 3x)dy\ dx = 0$

5. $\displaystyle\int_0^{\pi/2} \int_0^{\pi/2} (-y\cos x + x\sin y)dy\ dx = 0$

7. $\displaystyle\iint\limits_{R} [1 - (-1)]dA = 2\iint\limits_{R} dA = 8\pi$

9. $\displaystyle\iint\limits_{R} \left(-\frac{y}{1+y} - \frac{1}{1+y}\right)dA = -\iint\limits_{R} dA = -4$

11. $\displaystyle\iint\limits_{R} \left(-\frac{y^2}{1+y^2} - \frac{1}{1+y^2}\right)dA = -\iint\limits_{R} dA = -1$

13. $\displaystyle\int_0^1 \int_{x^2}^{\sqrt{x}} (y^2 - x^2)dy\ dx = 0$

15. **(a)** $C : x = \cos t, y = \sin t, 0 \le t \le 2\pi$;

$$\oint_C = \int_0^{2\pi} \left(e^{\sin t}(-\sin t) + \sin t\cos te^{\cos t}\right)dt \approx -3.550999378;$$

$$\iint\limits_{R} \left[\frac{\partial}{\partial x}(ye^x) - \frac{\partial}{\partial y}e^y\right]dA = \iint\limits_{R} [ye^x - e^y]\,dA$$

$$= \int_0^{2\pi} \int_0^1 \left[r\sin\theta e^{r\cos\theta} - e^{r\sin\theta}\right]r\,dr\,d\theta \approx -3.550999378$$

(b) $C_1 : x = t, y = t^2, 0 \le t \le 1; \displaystyle\int_{C_1} [e^y\ dx + ye^x\ dy] = \int_0^1 \left[e^{t^2} + 2t^3e^t\right]dt \approx 2.589524432$

$C_2 : x = t^2, y = t, 0 \le t \le 1; \displaystyle\int_{C_2} [e^y\ dx + ye^x\ dy] = \int_0^1 \left[2te^t + te^{t^2}\right]dt = \dfrac{e+3}{2} \approx 2.859140914$

$$\int_{C_1} - \int_{C_2} \approx -0.269616482; \quad \iint_R = \int_0^1 \int_{x^2}^{\sqrt{x}} [ye^x - e^y]\, dy\, dx \approx -0.269616482$$

17. $A = \dfrac{1}{2}\oint_C -y\,dx + x\,dy = \dfrac{1}{2}\int_0^{2\pi}(3a^2\sin^4\phi\cos^2\phi + 3a^2\cos^4\phi\sin^2\phi)\,d\phi$

$$= \dfrac{3}{2}a^2\int_0^{2\pi}\sin^2\phi\cos^2\phi\,d\phi = \dfrac{3}{8}a^2\int_0^{2\pi}\sin^2 2\phi\,d\phi = 3\pi a^2/8$$

19. $C_1 : (0,0)$ to $(a,0)$; $x = at, y = 0, 0 \le t \le 1$

$C_2 : (a,0)$ to $(a\cos t_0, b\sin t_0)$; $x = a\cos t, y = b\sin t, 0 \le t \le t_0$

$C_3 : (a\cos t_0, b\sin t_0)$ to $(0,0)$; $x = -a(\cos t_0)t, y = -b(\sin t_0)t, -1 \le t \le 0$

$$A = \dfrac{1}{2}\oint_C -y\,dx + x\,dy = \dfrac{1}{2}\int_0^1 (0)\,dt + \dfrac{1}{2}\int_0^{t_0} ab\,dt + \dfrac{1}{2}\int_{-1}^0 (0)\,dt = \dfrac{1}{2}ab\,t_0$$

21. $\operatorname{curl}\mathbf{F}(x,y,z) = -g_z\mathbf{i} + f_z\mathbf{j} + (g_x - f_y)\mathbf{k} = f_z\mathbf{j}(g_x - f_y)\mathbf{k}$, since f and g are independent of z. Thus

$$\iint_R \operatorname{curl}\mathbf{F}\cdot\mathbf{k}\,dA = \iint_R (f_x - g_y)\,dA = \int_C f(x,y)\,dx + g(x,y)\,dy = \int_C \mathbf{F}\cdot d\mathbf{r} \text{ by Green's Theorem.}$$

23. Let C_1 denote the graph of $g(x)$ from left to right, and C_2 the graph of $f(x)$ from left to right. On the vertical sides $x = \text{const}$, and so $dx = 0$ there. Thus the area between the curves is

$$A(R) = \iint_R dA = -\int_C y\,dx = -\int_{C_1} g(x)\,dx + \int_{C_2} f(x)\,dx$$

$$= -\int_a^b g(x)\,dx + \int_a^b f(x)\,dx = \int_a^b (f(x) - g(x))\,dx$$

25. $W = \displaystyle\iint_R y\,dA = \int_0^{\pi}\int_0^5 r^2\sin\theta\,dr\,d\theta = 250/3$

27. $\displaystyle\oint_C y\,dx - x\,dy = \iint_R (-2)dA = -2\int_0^{2\pi}\int_0^{a(1+\cos\theta)} r\,dr\,d\theta = -3\pi a^2$

29. $A = \displaystyle\int_0^1\int_{x^3}^x dy\,dx = \dfrac{1}{4};\ C_1 : x = t, y = t^3, 0 \le t \le 1, \int_{C_1} x^2\,dy = \int_0^1 t^2(3t^2)\,dt = \dfrac{3}{5}$

$C_2 : x = t, y = t, 0 \le t \le 1; \displaystyle\int_{C_2} x^2\,dy = \int_0^1 t^2\,dt = \dfrac{1}{3}, \oint_C x^2\,dy = \int_{C_1} - \int_{C_2} = \dfrac{3}{5} - \dfrac{1}{3} = \dfrac{4}{15}, \bar{x} = \dfrac{8}{15}$

$\displaystyle\int_C y^2\,dx = \int_0^1 t^6\,dt - \int_0^1 t^2\,dt = \dfrac{1}{7} - \dfrac{1}{3} = -\dfrac{4}{21}, \bar{y} = \dfrac{8}{21}, \text{ centroid } \left(\dfrac{8}{15}, \dfrac{8}{21}\right)$

31. $\bar{x} = 0$ from the symmetry of the region,

$C_1 : (a,0)$ to $(-a,0)$ along $y = \sqrt{a^2 - x^2}$; $x = a\cos t, y = a\sin t, 0 \le t \le \pi$

$C_2 : (-a,0)$ to $(a,0)$; $x = t, y = 0, -a \le t \le a$

$$A = \pi a^2/2, \quad \bar{y} = -\frac{1}{2A}\left[\int_0^\pi -a^3\sin^3 t\,dt + \int_{-a}^a (0)dt\right]$$

$$= -\frac{1}{\pi a^2}\left(-\frac{4a^3}{3}\right) = \frac{4a}{3\pi}; \text{ centroid } \left(0, \frac{4a}{3\pi}\right)$$

33. From Green's Theorem, the given integral equals $\iint_R (1-x^2-y^2)dA$ where R is the region enclosed by C. The value of this integral is maximum if the integration extends over the largest region for which the integrand $1-x^2-y^2$ is nonnegative so we want $1-x^2-y^2 \geq 0$, $x^2+y^2 \leq 1$. The largest region is that bounded by the circle $x^2+y^2 = 1$ which is the desired curve C.

35. $\int_C \mathbf{F}\cdot d\mathbf{r} = \int_C (x^2+y)\,dx + (4x - \cos y)\,dy = 3\iint_R dA = 3(25-2) = 69$

EXERCISE SET 16.5

1. R is the annular region between $x^2+y^2 = 1$ and $x^2+y^2 = 4$;

$$\iint_\sigma z^2 dS = \iint_R (x^2+y^2)\sqrt{\frac{x^2}{x^2+y^2} + \frac{y^2}{x^2+y^2} + 1}\,dA$$

$$= \sqrt{2}\iint_R (x^2+y^2)dA = \sqrt{2}\int_0^{2\pi}\int_1^2 r^3\,dr\,d\theta = \frac{15}{2}\pi\sqrt{2}.$$

3. Let $\mathbf{r}(u,v) = \cos u\mathbf{i} + v\mathbf{j} + \sin u\mathbf{k}, 0 \leq u \leq \pi, 0 \leq v \leq 1$. Then $\mathbf{r}_u = -\sin u\mathbf{i} + \cos u\mathbf{k}, \mathbf{r}_v = \mathbf{j}$,

$$\mathbf{r}_u \times \mathbf{r}_v = -\cos u\mathbf{i} - \sin u\mathbf{k}, \|\mathbf{r}_u \times \mathbf{r}_v\| = 1, \iint_\sigma x^2 y\,dS = \int_0^1\int_0^\pi v\cos^2 u\,du\,dv = \pi/4$$

5. If we use the projection of σ onto the xz-plane then $y = 1-x$ and R is the rectangular region in the xz-plane enclosed by $x = 0$, $x = 1$, $z = 0$ and $z = 1$;

$$\iint_\sigma (x-y-z)dS = \iint_R (2x-1-z)\sqrt{2}dA = \sqrt{2}\int_0^1\int_0^1 (2x-1-z)dz\,dx = -\sqrt{2}/2$$

7. There are six surfaces, parametrized by projecting onto planes:

$\sigma_1 : z = 0; 0 \leq x \leq 1, 0 \leq y \leq 1$ (onto xy-plane), $\sigma_2 : x = 0; 0 \leq y \leq 1, 0 \leq z \leq 1$ (onto yz-plane),
$\sigma_3 : y = 0; 0 \leq x \leq 1, 0 \leq z \leq 1$ (onto xz-plane), $\sigma_4 : z = 1; 0 \leq x \leq 1, 0 \leq y \leq 1$ (onto xy-plane),
$\sigma_5 : x = 1; 0 \leq y \leq 1, 0 \leq z \leq 1$ (onto yz-plane), $\sigma_6 : y = 1; 0 \leq x \leq 1, 0 \leq z \leq 1$ (onto xz-plane).

By symmetry the integrals over σ_1, σ_2 and σ_3 are equal, as are those over σ_4, σ_5 and σ_6, and

$$\iint_{\sigma_1}(x+y+z)dS = \int_0^1\int_0^1 (x+y)dx\,dy = 1; \quad \iint_{\sigma_4}(x+y+z)dS = \int_0^1\int_0^1 (x+y+1)dx\,dy = 2,$$

thus, $\iint_\sigma (x+y+z)dS = 3\cdot 1 + 3\cdot 2 = 9.$

9. **(a)** The integral is improper because the function $z(x, y)$ is not differentiable when $x^2 + y^2 = 1$.

   (b) Fix r_0, $0 < r_0 < 1$. Then $z + 1 = \sqrt{1 - x^2 - y^2} + 1$, and

   $$\iint_{\sigma_{r0}} (z + 1)\, dS = \iint_{\sigma_{r0}} (\sqrt{1 - {}^{\backprime}x^2 - y^2} + 1)\sqrt{1 + \frac{x^2}{1 - x^2 - y^2} + \frac{y^2}{1 - x^2 - y^2}}\, dx\, dy$$

   $$= \int_0^{2\pi} \int_0^{r_0} (\sqrt{1 - r^2} + 1)\frac{1}{\sqrt{1 - r^2}} r\, dr\, d\theta = 2\pi\left(1 + \frac{1}{2}r_0^2 - \sqrt{1 - r_0^2}\right), \text{ and, after passing to}$$

   the limit as $r_0 \to 1$, $\displaystyle\iint_\sigma (z + 1)\, dS = 3\pi$

   (c) Let $\mathbf{r}(\phi, \theta) = \sin\phi\cos\theta\mathbf{i} + \sin\phi\sin\theta\mathbf{j} + \cos\phi\mathbf{k}, 0 \le \theta \le 2\pi, 0 \le \phi \le \pi/2; \|\mathbf{r}_\phi \times \mathbf{r}_\theta\| = \sin\phi$,

   $$\iint_\sigma (1 + \cos\phi)\, dS = \int_0^{2\pi} \int_0^{\pi/2} (1 + \cos\phi)\sin\phi\, d\phi\, d\theta$$

   $$= 2\pi \int_0^{\pi/2} (1 + \cos\phi)\sin\phi\, d\phi = 3\pi$$

11. **(a)** Subdivide the right hemisphere $\sigma \cap \{x > 0\}$ into patches, each patch being as small as desired (i). For each patch there is a corresponding patch on the left hemisphere $\sigma \cap \{x < 0\}$ which is the reflection through the yz-plane. Condition (ii) now follows.

    (b) Use the patches in Part (a) and the function $f(x, y, z) = x^n$ to define the sum in Definition 16.5.1. The patches of the sum divide into two classes, each the negative of the other since n is odd. Thus the sum adds to zero. Since x^n is a continuous function the limit exists and must also be zero, $\int_\sigma x^n\, dS = 0$.

13. **(a)** Permuting the variables x, y, z by sending $x \to y \to z \to x$ will leave the integrals equal, through symmetry in the variables.

    (b) $\displaystyle\iint_\sigma (x^2 + y^2 + z^2)\, dS = $ surface area of sphere, so each integral contributes one third, i.e.

    $$\iint_\sigma x^2\, dS = \frac{1}{3}\left[\iint_\sigma x^2\, dS + \iint_\sigma y^2\, dS \iint_\sigma z^2\, dS\right].$$

    (c) Since σ has radius 1, $\displaystyle\iint_\sigma dS$ is the surface area of the sphere, which is 4π,

    therefore $\displaystyle\iint_\sigma x^2\, dS = \frac{4}{3}\pi$.

15. **(a)** $\displaystyle\frac{\sqrt{29}}{16} \int_0^6 \int_0^{(12-2x)/3} xy(12 - 2x - 3y)\, dy\, dx$

    (b) $\displaystyle\frac{\sqrt{29}}{4} \int_0^3 \int_0^{(12-4z)/3} yz(12 - 3y - 4z)\, dy\, dz$

    (c) $\displaystyle\frac{\sqrt{29}}{9} \int_0^3 \int_0^{6-2z} xz(12 - 2x - 4z)\, dx\, dz$

17. $18\sqrt{29}/5$

19. $\displaystyle\int_0^4 \int_1^2 y^3 z\sqrt{4y^2+1}\,dy\,dz;\ \frac{1}{2}\int_0^4 \int_1^4 xz\sqrt{1+4x}\,dx\,dz$

21. $391\sqrt{17}/15 - 5\sqrt{5}/3$

23. $z = \sqrt{4-x^2}, \dfrac{\partial z}{\partial x} = -\dfrac{x}{\sqrt{4-x^2}}, \dfrac{\partial z}{\partial y} = 0;$

$$\iint\limits_{\sigma} \delta_0\,dS = \delta_0 \iint\limits_{R} \sqrt{\frac{x^2}{4-x^2}+1}\,dA = 2\delta_0 \int_0^4 \int_0^1 \frac{1}{\sqrt{4-x^2}}\,dx\,dy = \frac{4}{3}\pi\delta_0.$$

25. $z = 4 - y^2$, R is the rectangular region enclosed by $x = 0$, $x = 3$, $y = 0$ and $y = 3$;

$$\iint\limits_{\sigma} y\,dS = \iint\limits_{R} y\sqrt{4y^2+1}\,dA = \int_0^3 \int_0^3 y\sqrt{4y^2+1}\,dy\,dx = \frac{1}{4}(37\sqrt{37}-1).$$

27. $\displaystyle M = \iint\limits_{\sigma} \delta(x,y,z)\,dS = \iint\limits_{\sigma} \delta_0\,dS = \delta_0 \iint\limits_{\sigma} dS = \delta_0 S$

29. By symmetry $\bar{x} = \bar{y} = 0$.

$$\iint\limits_{\sigma} dS = \iint\limits_{R} \sqrt{x^2+y^2+1}\,dA = \int_0^{2\pi} \int_0^{\sqrt{8}} \sqrt{r^2+1}\,r\,dr\,d\theta = \frac{52\pi}{3},$$

$$\iint\limits_{\sigma} z\,dS = \iint\limits_{R} z\sqrt{x^2+y^2+1}\,dA = \frac{1}{2}\iint\limits_{R}(x^2+y^2)\sqrt{x^2+y^2+1}\,dA$$

$$= \frac{1}{2}\int_0^{2\pi} \int_0^{\sqrt{8}} r^3\sqrt{r^2+1}\,dr\,d\theta = \frac{596\pi}{15}$$

so $\bar{z} = \dfrac{596\pi/15}{52\pi/3} = \dfrac{149}{65}$. The centroid is $(\bar{x},\bar{y},\bar{z}) = (0,0,149/65)$.

31. $\partial\mathbf{r}/\partial u = \cos v\,\mathbf{i} + \sin v\,\mathbf{j} + 3\mathbf{k}, \partial\mathbf{r}/\partial v = -u\sin v\,\mathbf{i} + u\cos v\,\mathbf{j}, \|\partial\mathbf{r}/\partial u \times \partial\mathbf{r}/\partial v\| = \sqrt{10}\,u;$

$$3\sqrt{10}\iint\limits_{R} u^4 \sin v \cos v\,dA = 3\sqrt{10}\int_0^{\pi/2} \int_1^2 u^4 \sin v \cos v\,du\,dv = 93/\sqrt{10}$$

33. $\partial\mathbf{r}/\partial u = \cos v\,\mathbf{i} + \sin v\,\mathbf{j} + 2u\mathbf{k}, \partial\mathbf{r}/\partial v = -u\sin v\,\mathbf{i} + u\cos v\,\mathbf{j}, \|\partial\mathbf{r}/\partial u \times \partial\mathbf{r}/\partial v\| = u\sqrt{4u^2+1};$

$$\iint\limits_{R} u\,dA = \int_0^{\pi} \int_0^{\sin v} u\,du\,dv = \pi/4$$

35. $\partial z/\partial x = -2xe^{-x^2-y^2}, \partial z/\partial y = -2ye^{-x^2-y^2},$

$(\partial z/\partial x)^2 + (\partial z/\partial y)^2 + 1 = 4(x^2+y^2)e^{-2(x^2+y^2)} + 1$; use polar coordinates to get

$$M = \int_0^{2\pi} \int_0^3 r^2\sqrt{4r^2 e^{-2r^2}+1}\,dr\,d\theta \approx 57.895751$$

EXERCISE SET 16.6

1. (a) zero (b) zero (c) positive
   (d) negative (e) zero (f) zero

3. (a) positive (b) zero (c) positive
   (d) zero (e) positive (f) zero

5. (a) $\mathbf{n} = -\cos v\mathbf{i} - \sin v\mathbf{j}$ (b) inward, by inspection

7. $\mathbf{n} = -z_x\mathbf{i} - z_y\mathbf{j} + \mathbf{k}$, $\displaystyle\iint\limits_R \mathbf{F}\cdot\mathbf{n}\,dS = \iint\limits_R (2x^2 + 2y^2 + 2(1 - x^2 - y^2))\,dS = \int_0^{2\pi}\int_0^1 2r\,dr\,d\theta = 2\pi$

9. R is the annular region enclosed by $x^2 + y^2 = 1$ and $x^2 + y^2 = 4$;

$$\iint\limits_\sigma \mathbf{F}\cdot\mathbf{n}\,dS = \iint\limits_R \left(-\frac{x^2}{\sqrt{x^2 + y^2}} - \frac{y^2}{\sqrt{x^2 + y^2}} + 2z\right)dA$$

$$= \iint\limits_R \sqrt{x^2 + y^2}\,dA = \int_0^{2\pi}\int_1^2 r^2\,dr\,d\theta = \frac{14\pi}{3}.$$

11. R is the circular region enclosed by $x^2 + y^2 - y = 0$; $\displaystyle\iint\limits_\sigma \mathbf{F}\cdot\mathbf{n}\,dS = \iint\limits_R (-x)dA = 0$ since the region R is symmetric across the y-axis.

13. $\partial\mathbf{r}/\partial u = \cos v\mathbf{i} + \sin v\mathbf{j} - 2u\mathbf{k}$, $\partial\mathbf{r}/\partial v = -u\sin v\mathbf{i} + u\cos v\mathbf{j}$,
$\partial\mathbf{r}/\partial u \times \partial\mathbf{r}/\partial v = 2u^2\cos v\mathbf{i} + 2u^2\sin v\mathbf{j} + u\mathbf{k}$;

$$\iint\limits_R (2u^3 + u)\,dA = \int_0^{2\pi}\int_1^2 (2u^3 + u)du\,dv = 18\pi$$

15. $\partial\mathbf{r}/\partial u = \cos v\mathbf{i} + \sin v\mathbf{j} + 2\mathbf{k}$, $\partial\mathbf{r}/\partial v = -u\sin v\mathbf{i} + u\cos v\mathbf{j}$,
$\partial\mathbf{r}/\partial u \times \partial\mathbf{r}/\partial v = -2u\cos v\mathbf{i} - 2u\sin v\mathbf{j} + u\mathbf{k}$;

$$\iint\limits_R u^2\,dA = \int_0^\pi\int_0^{\sin v} u^2\,du\,dv = 4/9$$

17. In each part, divide σ into the six surfaces
$\sigma_1 : x = -1$ with $|y| \le 1$, $|z| \le 1$, and $\mathbf{n} = -\mathbf{i}$, $\sigma_2 : x = 1$ with $|y| \le 1$, $|z| \le 1$, and $\mathbf{n} = \mathbf{i}$,
$\sigma_3 : y = -1$ with $|x| \le 1$, $|z| \le 1$, and $\mathbf{n} = -\mathbf{j}$, $\sigma_4 : y = 1$ with $|x| \le 1$, $|z| \le 1$, and $\mathbf{n} = \mathbf{j}$,
$\sigma_5 : z = -1$ with $|x| \le 1$, $|y| \le 1$, and $\mathbf{n} = -\mathbf{k}$, $\sigma_6 : z = 1$ with $|x| \le 1$, $|y| \le 1$, and $\mathbf{n} = \mathbf{k}$,

(a) $\displaystyle\iint\limits_{\sigma_1}\mathbf{F}\cdot\mathbf{n}\,dS = \iint\limits_{\sigma_1} dS = 4$, $\displaystyle\iint\limits_{\sigma_2}\mathbf{F}\cdot\mathbf{n}\,dS = \iint\limits_{\sigma_2} dS = 4$, and $\displaystyle\iint\limits_{\sigma_i}\mathbf{F}\cdot\mathbf{n}\,dS = 0$ for

$i = 3, 4, 5, 6$ so $\displaystyle\iint\limits_\sigma \mathbf{F}\cdot\mathbf{n}\,dS = 4 + 4 + 0 + 0 + 0 + 0 = 8.$

(b) $\iint\limits_{\sigma_1} \mathbf{F} \cdot \mathbf{n}\, dS = \iint\limits_{\sigma_1} dS = 4$, similarly $\iint\limits_{\sigma_i} \mathbf{F} \cdot \mathbf{n}\, dS = 4$ for $i = 2, 3, 4, 5, 6$ so

$\iint\limits_{\sigma} \mathbf{F} \cdot \mathbf{n}\, dS = 4 + 4 + 4 + 4 + 4 + 4 = 24.$

(c) $\iint\limits_{\sigma_1} \mathbf{F} \cdot \mathbf{n}\, dS = -\iint\limits_{\sigma_1} dS = -4$, $\iint\limits_{\sigma_2} \mathbf{F} \cdot \mathbf{n}\, dS = 4$, similarly $\iint\limits_{\sigma_i} \mathbf{F} \cdot \mathbf{n}\, dS = -4$ for $i = 3, 5$

and $\iint\limits_{\sigma_i} \mathbf{F} \cdot \mathbf{n}\, dS = 4$ for $i = 4, 6$ so $\iint\limits_{\sigma} \mathbf{F} \cdot \mathbf{n}\, dS = -4 + 4 - 4 + 4 - 4 + 4 = 0.$

19. R is the circular region enclosed by $x^2 + y^2 = 1$; $x = r\cos\theta, y = r\sin\theta, z = r$,

$\mathbf{n} = \cos\theta\,\mathbf{i} + \sin\theta\,\mathbf{j} - \mathbf{k}$;

$\iint\limits_{\sigma} \mathbf{F} \cdot \mathbf{n}\, dS = \iint\limits_{R} (\cos\theta + \sin\theta - 1)\, dA = \int_0^{2\pi} \int_0^1 (\cos\theta + \sin\theta - 1)\, r\, dr\, d\theta = -\pi.$

21. (a) $\mathbf{n} = \dfrac{1}{\sqrt{3}}[\mathbf{i} + \mathbf{j} + \mathbf{k}]$,

$V = \int_{\sigma} \mathbf{F} \cdot \mathbf{n}\, dS = \int_0^1 \int_0^{1-x} (2x - 3y + 1 - x - y)\, dy\, dx = 0 \text{ m}^3$

(b) $m = 0 \cdot 806 = 0$ kg

23. (a) $G(x, y, z) = x - g(y, z)$, $\nabla G = \mathbf{i} - \dfrac{\partial g}{\partial y}\mathbf{j} - \dfrac{\partial g}{\partial z}\mathbf{k}$, apply Theorem 16.6.3:

$\iint\limits_{\sigma} \mathbf{F} \cdot \mathbf{n} dS = \iint\limits_{R} \mathbf{F} \cdot \left(\mathbf{i} - \dfrac{\partial x}{\partial y}\mathbf{j} - \dfrac{\partial x}{\partial z}\mathbf{k}\right) dA$, if σ is oriented by front normals, and

$\iint\limits_{\sigma} \mathbf{F} \cdot \mathbf{n} dS = \iint\limits_{R} \mathbf{F} \cdot \left(-\mathbf{i} + \dfrac{\partial x}{\partial y}\mathbf{j} + \dfrac{\partial x}{\partial z}\mathbf{k}\right) dA$, if σ is oriented by back normals,

where R is the projection of σ onto the yz-plane.

(b) R is the semicircular region in the yz-plane enclosed by $z = \sqrt{1 - y^2}$ and $z = 0$;

$\iint\limits_{\sigma} \mathbf{F} \cdot \mathbf{n}\, dS = \iint\limits_{R} (-y - 2yz + 16z)\, dA = \int_{-1}^{1} \int_0^{\sqrt{1-y^2}} (-y - 2yz + 16z)\, dz\, dy = \dfrac{32}{3}.$

25. (a) On the sphere, $\|\mathbf{r}\| = a$ so $\mathbf{F} = a^k\mathbf{r}$ and $\mathbf{F} \cdot \mathbf{n} = a^k\mathbf{r} \cdot (\mathbf{r}/a) = a^{k-1}\|\mathbf{r}\|^2 = a^{k-1}a^2 = a^{k+1}$,

hence $\iint\limits_{\sigma} \mathbf{F} \cdot \mathbf{n}\, dS = a^{k+1} \iint\limits_{\sigma} dS = a^{k+1}(4\pi a^2) = 4\pi a^{k+3}.$

(b) If $k = -3$, then $\iint\limits_{\sigma} \mathbf{F} \cdot \mathbf{n}\, dS = 4\pi.$

EXERCISE SET 16.7

1. $\sigma_1 : x = 0, \mathbf{F} \cdot \mathbf{n} = -x = 0, \displaystyle\iint_{\sigma_1} (0) dA = 0$ $\qquad$ $\sigma_2 : x = 1, \mathbf{F} \cdot \mathbf{n} = x = 1, \displaystyle\iint_{\sigma_2} (1) dA = 1$

$\sigma_3 : y = 0, \mathbf{F} \cdot \mathbf{n} = -y = 0, \displaystyle\iint_{\sigma_3} (0) dA = 0$ $\qquad$ $\sigma_4 : y = 1, \mathbf{F} \cdot \mathbf{n} = y = 1, \displaystyle\iint_{\sigma_4} (1) dA = 1$

$\sigma_5 : z = 0, \mathbf{F} \cdot \mathbf{n} = -z = 0, \displaystyle\iint_{\sigma_5} (0) dA = 0$ $\qquad$ $\sigma_6 : z = 1, \mathbf{F} \cdot \mathbf{n} = z = 1, \displaystyle\iint_{\sigma_6} (1) dA = 1$

$$\iint_{\sigma} \mathbf{F} \cdot \mathbf{n} = 3; \quad \iiint_{G} \text{div } \mathbf{F} dV = \iiint_{G} 3 dV = 3$$

3. $\sigma_1 : z = 1, \mathbf{n} = \mathbf{k}, \mathbf{F} \cdot \mathbf{n} = z^2 = 1, \displaystyle\iint_{\sigma_1} (1) dS = \pi,$

$\sigma_2 : \mathbf{n} = 2x\mathbf{i} + 2y\mathbf{j} - \mathbf{k}, \mathbf{F} \cdot \mathbf{n} = 4x^2 - 4x^2 y^2 - x^4 - 3y^4,$

$$\iint_{\sigma_2} \mathbf{F} \cdot \mathbf{n} \, dS = \int_0^{2\pi} \int_0^1 \left[4r^2 \cos^2\theta - 4r^4 \cos^2\theta \sin^2\theta - r^4 \cos^4\theta - 3r^4 \sin^4\theta \right] r \, dr \, d\theta = \frac{\pi}{3};$$

$$\iint_{\sigma} = \frac{4\pi}{3}$$

$$\iiint_{G} \text{div } \mathbf{F} dV = \iiint_{G} (2 + z) dV = \int_0^{2\pi} \int_0^1 \int_{r^2}^1 (2 + z) dz \, r \, dr \, d\theta = 4\pi/3$$

5. G is the rectangular solid; $\displaystyle\iiint_{G} \text{div } \mathbf{F} \, dV = \int_0^2 \int_0^1 \int_0^3 (2x - 1) \, dx \, dy \, dz = 12.$

7. G is the cylindrical solid;

$$\iiint_{G} \text{div } \mathbf{F} \, dV = 3 \iiint_{G} dV = (3)(\text{volume of cylinder}) = (3)[\pi a^2(1)] = 3\pi a^2.$$

9. G is the cylindrical solid;

$$\iiint_{G} \text{div } \mathbf{F} \, dV = 3 \iiint_{G} (x^2 + y^2 + z^2) dV = 3 \int_0^{2\pi} \int_0^2 \int_0^3 (r^2 + z^2) r \, dz \, dr \, d\theta = 180\pi.$$

11. G is the hemispherical solid bounded by $z = \sqrt{4 - x^2 - y^2}$ and the xy-plane;

$$\iiint_{G} \text{div } \mathbf{F} \, dV = 3 \iiint_{G} (x^2 + y^2 + z^2) dV = 3 \int_0^{2\pi} \int_0^{\pi/2} \int_0^2 \rho^4 \sin\phi \, d\rho \, d\phi \, d\theta = \frac{192\pi}{5}.$$

13. G is the conical solid;

$$\iiint_{G} \text{div } \mathbf{F} \, dV = 2 \iiint_{G} (x + y + z) dV = 2 \int_0^{2\pi} \int_0^1 \int_r^1 (r\cos\theta + r\sin\theta + z) r \, dz \, dr \, d\theta = \frac{\pi}{2}.$$

15. G is the solid bounded by $z = 4 - x^2$, $y + z = 5$, and the coordinate planes;

$$\iiint\limits_{G} \text{div } \mathbf{F} \, dV = 4 \iiint\limits_{G} x^2 dV = 4 \int_{-2}^{2} \int_{0}^{4-x^2} \int_{0}^{5-z} x^2 dy \, dz \, dx = \frac{4608}{35}.$$

17. $\iint\limits_{\sigma} \mathbf{F} \cdot \mathbf{n} \, dS = 3[\pi(3^2)(5)] = 135\pi$

19. **(a)** $\mathbf{F} = x\mathbf{i} + y\mathbf{j} + z\mathbf{k}$, div $\mathbf{F} = 3$ **(b)** $\mathbf{F} = -x\mathbf{i} - y\mathbf{j} - z\mathbf{k}$, div $\mathbf{F} = -3$

21. $0 = \iiint\limits_{R} \mathbf{F} \text{ div } dV = \iint\limits_{\sigma} \mathbf{F} \cdot \mathbf{n} \, dS$. Let σ_1 denote that part of σ on which $\mathbf{F} \cdot \mathbf{n} > 0$ and let σ_2 denote the part where $\mathbf{F} \cdot \mathbf{n} < 0$. If $\iint\limits_{\sigma_1} \mathbf{F} \cdot \mathbf{n} > 0$ then the integral over σ_2 is negative (and equal in magnitude). Thus the boundary between σ_1 and σ_2 is infinite, hence $\mathbf{F}$ and $\mathbf{n}$ are perpendicular on an infinite set.

23. $\iint\limits_{\sigma} \text{curl } \mathbf{F} \cdot \mathbf{n} \, dS = \iiint\limits_{G} \text{div(curl } \mathbf{F})dV = \iiint\limits_{G} (0)dV = 0$

25. $\iint\limits_{\sigma} (f\nabla g) \cdot \mathbf{n} = \iiint\limits_{G} \text{div } (f\nabla g)dV = \iiint\limits_{G} (f\nabla^2 g + \nabla f \cdot \nabla g)dV$ by Exercise 31, Section 16.1.

27. Since $\mathbf{v}$ is constant, $\nabla \cdot \mathbf{v} = 0$. Let $\mathbf{F} = f\mathbf{v}$; then div$\mathbf{F} = (\nabla f)\mathbf{v}$ and by the Divergence Theorem
$$\iint\limits_{\sigma} f\mathbf{v} \cdot \mathbf{n} \, dS = \iint\limits_{\sigma} \mathbf{F} \cdot \mathbf{n} \, dS = \iiint\limits_{G} \text{div}\mathbf{F} \, dV = \iiint\limits_{G} (\nabla f) \cdot \mathbf{v} \, dV$$

29. div $\mathbf{F} = 0$; no sources or sinks.

31. div $\mathbf{F} = 3x^2 + 3y^2 + 3z^2$; sources at all points except the origin, no sinks.

33. Let σ_1 be the portion of the paraboloid $z = 1 - x^2 - y^2$ for $z \geq 0$, and σ_2 the portion of the plane $z = 0$ for $x^2 + y^2 \leq 1$. Then
$$\iint\limits_{\sigma_1} \mathbf{F} \cdot \mathbf{n} \, dS = \iint\limits_{R} \mathbf{F} \cdot (2x\mathbf{i} + 2y\mathbf{j} + \mathbf{k}) \, dA$$
$$= \int_{-1}^{1} \int_{-\sqrt{1-x^2}}^{\sqrt{1-x^2}} (2x[x^2 y - (1 - x^2 - y^2)^2] + 2y(y^3 - x) + (2x + 2 - 3x^2 - 3y^2)) \, dy \, dx$$
$$= 3\pi/4;$$

$z = 0$ and $\mathbf{n} = -\mathbf{k}$ on σ_2 so $\mathbf{F} \cdot \mathbf{n} = 1 - 2x$, $\iint\limits_{\sigma_2} \mathbf{F} \cdot \mathbf{n} \, dS = \iint\limits_{\sigma_2} (1 - 2x)dS = \pi$. Thus

$\iint\limits_{\sigma} \mathbf{F} \cdot \mathbf{n} \, dS = 3\pi/4 + \pi = 7\pi/4$. But div $\mathbf{F} = 2xy + 3y^2 + 3$ so

$$\iiint\limits_{G} \text{div } \mathbf{F} \, dV = \int_{-1}^{1} \int_{-\sqrt{1-x^2}}^{\sqrt{1-x^2}} \int_{0}^{1-x^2-y^2} (2xy + 3y^2 + 3) \, dz \, dy \, dx = 7\pi/4.$$

EXERCISE SET 16.8

1. If σ is oriented with upward normals then C consists of three parts parametrized as
    $C_1 : \mathbf{r}(t) = (1-t)\mathbf{i} + t\mathbf{j}$ for $0 \le t \le 1$, $C_2 : \mathbf{r}(t) = (1-t)\mathbf{j} + t\mathbf{k}$ for $0 \le t \le 1$,
    $C_3 : \mathbf{r}(t) = t\mathbf{i} + (1-t)\mathbf{k}$ for $0 \le t \le 1$.

    $$\int_{C_1} \mathbf{F} \cdot d\mathbf{r} = \int_{C_2} \mathbf{F} \cdot d\mathbf{r} = \int_{C_3} \mathbf{F} \cdot d\mathbf{r} = \int_0^1 (3t-1)dt = \frac{1}{2} \text{ so}$$

    $$\oint_C \mathbf{F} \cdot d\mathbf{r} = \frac{1}{2} + \frac{1}{2} + \frac{1}{2} = \frac{3}{2}. \text{ curl } \mathbf{F} = \mathbf{i} + \mathbf{j} + \mathbf{k}, \ z = 1 - x - y, \ R \text{ is the triangular region in}$$

    the xy-plane enclosed by $x + y = 1$, $x = 0$, and $y = 0$;

    $$\iint_\sigma (\text{curl } \mathbf{F}) \cdot \mathbf{n} \, dS = 3 \iint_R dA = (3)(\text{area of } R) = (3) \left[\frac{1}{2}(1)(1) \right] = \frac{3}{2}.$$

3. If σ is oriented with upward normals then C can be parametrized as $\mathbf{r}(t) = a \cos t\mathbf{i} + a \sin t\mathbf{j}$ for
    $0 \le t \le 2\pi$.

    $$\oint_C \mathbf{F} \cdot d\mathbf{r} = \int_0^{2\pi} 0 \, dt = 0; \text{ curl } \mathbf{F} = \mathbf{0} \text{ so } \iint_\sigma (\text{curl } \mathbf{F}) \cdot \mathbf{n} \, dS = \iint_\sigma 0 \, dS = 0.$$

5. Take σ as the part of the plane $z = 0$ for $x^2 + y^2 \le 1$ with $\mathbf{n} = \mathbf{k}$; curl $\mathbf{F} = -3y^2\mathbf{i} + 2z\mathbf{j} + 2\mathbf{k}$,

    $$\iint_\sigma (\text{curl } \mathbf{F}) \cdot \mathbf{n} \, dS = 2 \iint_\sigma dS = (2)(\text{area of circle}) = (2)[\pi(1)^2] = 2\pi.$$

7. C is the boundary of R and curl $\mathbf{F} = 2\mathbf{i} + 3\mathbf{j} + 4\mathbf{k}$, so

    $$\oint \mathbf{F} \cdot \mathbf{r} = \iint_R \text{curl } \mathbf{F} \cdot \mathbf{n} \, dS = \iint_R 4 \, dA = 4(\text{area of } R) = 16\pi$$

9. curl $\mathbf{F} = x\mathbf{k}$, take σ as part of the plane $z = y$ oriented with upward normals, R is the circular
    region in the xy-plane enclosed by $x^2 + y^2 - y = 0$;

    $$\iint_\sigma (\text{curl } \mathbf{F}) \cdot \mathbf{n} \, dS = \iint_R x \, dA = \int_0^\pi \int_0^{\sin \theta} r^2 \cos \theta \, dr \, d\theta = 0.$$

11. curl $\mathbf{F} = \mathbf{i} + \mathbf{j} + \mathbf{k}$, take σ as the part of the plane $z = 0$ with $x^2 + y^2 \le a^2$ and $\mathbf{n} = \mathbf{k}$;

    $$\iint_\sigma (\text{curl } \mathbf{F}) \cdot \mathbf{n} \, dS = \iint_\sigma dS = \text{ area of circle } = \pi a^2.$$

13. (a) Take σ as the part of the plane $2x + y + 2z = 2$ in the first octant, oriented with downward
        normals; curl $\mathbf{F} = -x\mathbf{i} + (y-1)\mathbf{j} - \mathbf{k}$,

    $$\oint_C \mathbf{F} \cdot \mathbf{T} \, ds = \iint_\sigma (\text{curl } \mathbf{F}) \cdot \mathbf{n} \, dS$$

    $$= \iint_R \left(x - \frac{1}{2}y + \frac{3}{2} \right) dA = \int_0^1 \int_0^{2-2x} \left(x - \frac{1}{2}y + \frac{3}{2} \right) dy \, dx = \frac{3}{2}.$$

    (b) At the origin curl $\mathbf{F} = -\mathbf{j} - \mathbf{k}$ and with $\mathbf{n} = \mathbf{k}$, curl $\mathbf{F}(0,0,0) \cdot \mathbf{n} = (-\mathbf{j} - \mathbf{k}) \cdot \mathbf{k} = -1$.

(c) The rotation of $\mathbf{F}$ has its maximum value at the origin about the unit vector in the same direction as curl $\mathbf{F}(0,0,0)$ so $\mathbf{n} = -\dfrac{1}{\sqrt{2}}\mathbf{j} - \dfrac{1}{\sqrt{2}}\mathbf{k}$.

15. (a) The flow is independent of z and has no component in the direction of $\mathbf{k}$, and so by inspection the only nonzero component of the curl is in the direction of $\mathbf{k}$. However both sides of (9) are zero, as the flow is orthogonal to the curve C_a. Thus the curl is zero.

(b) Since the flow appears to be tangential to the curve C_a, it seems that the right hand side of (9) is nonzero, and thus the curl is nonzero, and points in the positive z-direction.

17. Since $\mathbf{F}$ is conservative, if C is any closed curve then $\displaystyle\int_C \mathbf{F} \cdot d\mathbf{r} = 0$. But $\displaystyle\int_C \mathbf{F} \cdot d\mathbf{r} = \int_C \mathbf{F} \cdot \mathbf{T}\, ds$ from (30) of Section 16.2. In equation (9) the direction of $\mathbf{n}$ is arbitrary, so for any fixed curve C_a the integral $\displaystyle\int_{C_a} \mathbf{F} \cdot \mathbf{T}\, ds = 0$. Thus curl $\mathbf{F}(P_0) \cdot \mathbf{n} = 0$. But $\mathbf{n}$ is arbitrary, so we conclude that curl $\mathbf{F} = \mathbf{0}$.

19. Parametrize C by $x = \cos t, y = \sin t, 0 \le t \le 2\pi$. But $\mathbf{F} = x^2 y\mathbf{i} + (y^3 - x)\mathbf{j} + (2x-1)\mathbf{k}$ along C so $\displaystyle\oint_C \mathbf{F} \cdot d\mathbf{r} = -5\pi/4$. Since curl $\mathbf{F} = (-2z-2)\mathbf{j} + (-1-x^2)\mathbf{k}$,

$$\iint\limits_{\sigma} (\text{curl } \mathbf{F}) \cdot \mathbf{n}\, dS = \iint\limits_{R} (\text{curl } \mathbf{F}) \cdot (2x\mathbf{i} + 2y\mathbf{j} + \mathbf{k})\, dA$$

$$= \int_{-1}^{1} \int_{-\sqrt{1-x^2}}^{\sqrt{1-x^2}} [2y(2x^2 + 2y^2 - 4) - 1 - x^2]\, dy\, dx = -5\pi/4$$

REVIEW EXERCISES, CHAPTER 16

3. $\mathbf{v} = (1-x)\mathbf{i} + (2-y)\mathbf{j}, \|\mathbf{v}\| = \sqrt{(1-x)^2 + (2-y)^2}$,

$\mathbf{u} = \dfrac{1}{\|\mathbf{v}\|}\mathbf{v} = \dfrac{1-x}{\sqrt{(1-x)^2 + (2-y)^2}}\mathbf{i} + \dfrac{2-y}{\sqrt{(1-x)^2 + (2-y)^2}}\mathbf{j}$

5. $\mathbf{i} + \mathbf{j} + \mathbf{k}$

7. (a) $\displaystyle\int_a^b \left[f(x(t), y(t))\dfrac{dx}{dt} + g(x(t), y(t))\dfrac{dy}{dt} \right] dt$

(b) $\displaystyle\int_a^b f(x(t), y(t))\sqrt{x'(t)^2 + y'(t)^2}\, dt$

11. $s = \theta, x = \cos\theta, y = \sin\theta, \displaystyle\int_0^{2\pi} (\cos\theta - \sin\theta)\, d\theta = 0$, also follows from odd function rule.

13. $\displaystyle\int_1^2 \left(\dfrac{t}{2t} - 2\dfrac{2t}{t} \right) dt = \int_1^2 \left(-\dfrac{7}{2} \right) dt = -\dfrac{7}{2}$

17. (a) If $h(x)\mathbf{F}$ is conservative, then $\dfrac{\partial}{\partial y}(yh(x)) = \dfrac{\partial}{\partial x}(-2xh(x))$, or $h(x) = -2h(x) - 2xh'(x)$ which has the general solution $x^3 h(x)^2 = C_1, h(x) = Cx^{-3/2}$, so $C\dfrac{y}{x^{3/2}}\mathbf{i} - C\dfrac{2}{x^{1/2}}\mathbf{j}$ is conservative, with potential function $\phi = -2Cy/\sqrt{x}$.

(b) If $g(y)\mathbf{F}(x,y)$ is conservative then $\dfrac{\partial}{\partial y}(yg(y)) = \dfrac{\partial}{\partial x}(-2xg(y))$, or $g(y) + yg'(y) = -2g(y)$, with general solution $g(y) = C/y^3$, so $\mathbf{F} = C\dfrac{1}{y^2}\mathbf{i} - C\dfrac{2x}{y^3}\mathbf{j}$ is conservative, with potential function Cx/y^2.

21. Let O be the origin, P the point with polar coordinates $\theta = \alpha, r = f(\alpha)$, and Q the point with polar coordinates $\theta = \beta, r = f(\beta)$. Let

$$C_1 : O \text{ to } P; \ x = t\cos\alpha, \ y = t\sin\alpha, \ 0 \le t \le f(\alpha), -y\frac{dx}{dt} + x\frac{dy}{dt} = 0$$

$$C_2 : P \text{ to } Q; \ x = f(t)\cos t, \ y = f(t)\sin t, \ \alpha \le \theta \le \beta, -y\frac{dx}{dt} + x\frac{dy}{dt} = f(t)^2$$

$$C_3 : Q \text{ to } O; \ x = -t\cos\beta, \ y = -t\sin\beta, \ -f(\beta) \le t \le 0, -y\frac{dx}{dt} + x\frac{dy}{dt} = 0$$

$$A = \frac{1}{2}\oint_C -y\,dx + x\,dy = \frac{1}{2}\int_\alpha^\beta [f(t)]^2\,dt; \text{ set } t = \theta \text{ and } r = f(\theta) = f(t), A = \frac{1}{2}\int_\alpha^\beta r^2\,d\theta.$$

23. $\displaystyle\iint_\sigma f(x,y,z)dS = \iint_R f(x(u,v),y(u,v),z(u,v))\|\mathbf{r}_u \times \mathbf{r}_v\|\,du\,dv$

25. Yes; by imagining a normal vector sliding around the surface it is evident that the surface has two sides.

27. $\mathbf{r} = x\mathbf{i} + y\mathbf{j} + (1 - x^2 - y^2)\mathbf{k}, \mathbf{r}_x \times \mathbf{r}_y = 2x\mathbf{i} + 2y\mathbf{j} + \mathbf{k}, \ \mathbf{F} = x\mathbf{i} + y\mathbf{i} + 2z\mathbf{k}$

$$\Phi = \iint_R \mathbf{F}\cdot(\mathbf{r}_x \times \mathbf{r}_y)\,dA = \iint_R (2x^2 + 2y^2 + 2(1 - x^2 - y^2))\,dA = 2A = 2\pi$$

31. By Exercise 30, $\displaystyle\iint_\sigma D_\mathbf{n}f\,dS = -\iiint_G [f_{xx} + f_{yy} + f_{zz}]\,dV = -6\iiint_G dV = -6\text{vol}(G) = -8\pi$

33. A computation of curl $\mathbf{F}$ shows that curl $\mathbf{F} = \mathbf{0}$ if and only if the three given equations hold. Moreover the equations hold if $\mathbf{F}$ is conservative, so it remains to show that $\mathbf{F}$ is conservative if curl $\mathbf{F} = \mathbf{0}$. Let C by any simple closed curve in the region. Since the region is simply connected, there is a piecewise smooth, oriented surface σ in the region with boundary C. By Stokes' Theorem,

$$\oint_C \mathbf{F}\cdot d\mathbf{r} = \iint_\sigma (\text{curl }\mathbf{F})\cdot\mathbf{n}\,dS = \iint_\sigma 0\,dS = 0.$$

By the 3-space analog of Theorem 16.3.2, $\mathbf{F}$ is conservative.

35. **(a)** conservative, $\phi(x,y,z) = -\cos x + yz$ **(b)** not conservative, $f_z \ne h_x$

APPENDIX A
Trigonometry Review

EXERCISE SET A

1. (a) $5\pi/12$ (b) $13\pi/6$ (c) $\pi/9$ (d) $23\pi/30$

3. (a) $12°$ (b) $(270/\pi)°$ (c) $288°$ (d) $540°$

5.

	$\sin\theta$	$\cos\theta$	$\tan\theta$	$\csc\theta$	$\sec\theta$	$\cot\theta$
(a)	$\sqrt{21}/5$	$2/5$	$\sqrt{21}/2$	$5/\sqrt{21}$	$5/2$	$2/\sqrt{21}$
(b)	$3/4$	$\sqrt{7}/4$	$3/\sqrt{7}$	$4/3$	$4/\sqrt{7}$	$\sqrt{7}/3$
(c)	$3/\sqrt{10}$	$1/\sqrt{10}$	3	$\sqrt{10}/3$	$\sqrt{10}$	$1/3$

7. $\sin\theta = 3/\sqrt{10}$, $\cos\theta = 1/\sqrt{10}$ **9.** $\tan\theta = \sqrt{21}/2$, $\csc\theta = 5/\sqrt{21}$

11. Let x be the length of the side adjacent to θ, then $\cos\theta = x/6 = 0.3$, $x = 1.8$.

13.

	θ	$\sin\theta$	$\cos\theta$	$\tan\theta$	$\csc\theta$	$\sec\theta$	$\cot\theta$
(a)	$225°$	$-1/\sqrt{2}$	$-1/\sqrt{2}$	1	$-\sqrt{2}$	$-\sqrt{2}$	1
(b)	$-210°$	$1/2$	$-\sqrt{3}/2$	$-1/\sqrt{3}$	2	$-2/\sqrt{3}$	$-\sqrt{3}$
(c)	$5\pi/3$	$-\sqrt{3}/2$	$1/2$	$-\sqrt{3}$	$-2/\sqrt{3}$	2	$-1/\sqrt{3}$
(d)	$-3\pi/2$	1	0	—	1	—	0

15.

	$\sin\theta$	$\cos\theta$	$\tan\theta$	$\csc\theta$	$\sec\theta$	$\cot\theta$
(a)	$4/5$	$3/5$	$4/3$	$5/4$	$5/3$	$3/4$
(b)	$-4/5$	$3/5$	$-4/3$	$-5/4$	$5/3$	$-3/4$
(c)	$1/2$	$-\sqrt{3}/2$	$-1/\sqrt{3}$	2	$-2\sqrt{3}$	$-\sqrt{3}$
(d)	$-1/2$	$\sqrt{3}/2$	$-1/\sqrt{3}$	-2	$2/\sqrt{3}$	$-\sqrt{3}$
(e)	$1/\sqrt{2}$	$1/\sqrt{2}$	1	$\sqrt{2}$	$\sqrt{2}$	1
(f)	$1/\sqrt{2}$	$-1/\sqrt{2}$	-1	$\sqrt{2}$	$-\sqrt{2}$	-1

17. (a) $x = 3\sin 25° \approx 1.2679$ (b) $x = 3/\tan(2\pi/9) \approx 3.5753$

19.

	$\sin\theta$	$\cos\theta$	$\tan\theta$	$\csc\theta$	$\sec\theta$	$\cot\theta$
(a)	$a/3$	$\sqrt{9-a^2}/3$	$a/\sqrt{9-a^2}$	$3/a$	$3/\sqrt{9-a^2}$	$\sqrt{9-a^2}/a$
(b)	$a/\sqrt{a^2+25}$	$5/\sqrt{a^2+25}$	$a/5$	$\sqrt{a^2+25}/a$	$\sqrt{a^2+25}/5$	$5/a$
(c)	$\sqrt{a^2-1}/a$	$1/a$	$\sqrt{a^2-1}$	$a/\sqrt{a^2-1}$	a	$1/\sqrt{a^2-1}$

21. (a) $\theta = 3\pi/4 \pm n\pi$, $n = 0, 1, 2, \ldots$
     (b) $\theta = \pi/3 \pm 2n\pi$ and $\theta = 5\pi/3 \pm 2n\pi$, $n = 0, 1, 2, \ldots$

23.   (a) $\theta = \pi/6 \pm n\pi$, $n = 0, 1, 2, \dots$
       (b) $\theta = 4\pi/3 \pm 2n\pi$ and $\theta = 5\pi/3 \pm 2n\pi$, $n = 0, 1, 2, \dots$

25.   (a) $\theta = 3\pi/4 \pm n\pi$, $n = 0, 1, 2, \dots$ (b) $\theta = \pi/6 \pm n\pi$, $n = 0, 1, 2, \dots$

27.   (a) $\theta = \pi/3 \pm 2n\pi$ and $\theta = 2\pi/3 \pm 2n\pi$, $n = 0, 1, 2, \dots$
       (b) $\theta = \pi/6 \pm 2n\pi$ and $\theta = 11\pi/6 \pm 2n\pi$, $n = 0, 1, 2, \dots$

29. $\sin\theta = 2/5$, $\cos\theta = -\sqrt{21}/5$, $\tan\theta = -2/\sqrt{21}$, $\csc\theta = 5/2$, $\sec\theta = -5/\sqrt{21}$, $\cot\theta = -\sqrt{21}/2$

31.   (a) $\theta = \pm n\pi$, $n = 0, 1, 2, \dots$ (b) $\theta = \pi/2 \pm n\pi$, $n = 0, 1, 2, \dots$
       (c) $\theta = \pm n\pi$, $n = 0, 1, 2, \dots$ (d) $\theta = \pm n\pi$, $n = 0, 1, 2, \dots$
       (e) $\theta = \pi/2 \pm n\pi$, $n = 0, 1, 2, \dots$ (f) $\theta = \pm n\pi$, $n = 0, 1, 2, \dots$

33.   (a) $s = r\theta = 4(\pi/6) = 2\pi/3$ cm   (b) $s = r\theta = 4(5\pi/6) = 10\pi/3$ cm

35. $\theta = s/r = 2/5$

37.   (a) $2\pi r = R(2\pi - \theta)$, $r = \dfrac{2\pi - \theta}{2\pi}R$

       (b) $h = \sqrt{R^2 - r^2} = \sqrt{R^2 - (2\pi - \theta)^2 R^2/(4\pi^2)} = \dfrac{\sqrt{4\pi\theta - \theta^2}}{2\pi}R$

39.   Let h be the altitude as shown in the figure, then
       $h = 3\sin 60° = 3\sqrt{3}/2$ so $A = \dfrac{1}{2}(3\sqrt{3}/2)(7) = 21\sqrt{3}/4.$

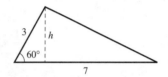

41.   Let x be the distance above the ground, then $x = 10\sin 67° \approx 9.2$ ft.

43.   From the figure, $h = x - y$ but $x = d\tan\beta$,
       $y = d\tan\alpha$ so $h = d(\tan\beta - \tan\alpha)$.

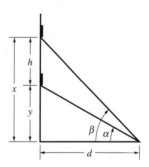

45.   (a) $\sin 2\theta = 2\sin\theta\cos\theta = 2(\sqrt{5}/3)(2/3) = 4\sqrt{5}/9$
       (b) $\cos 2\theta = 2\cos^2\theta - 1 = 2(2/3)^2 - 1 = -1/9$

47. $\sin 3\theta = \sin(2\theta + \theta) = \sin 2\theta\cos\theta + \cos 2\theta\sin\theta = (2\sin\theta\cos\theta)\cos\theta + (\cos^2\theta - \sin^2\theta)\sin\theta$
       $= 2\sin\theta\cos^2\theta + \sin\theta\cos^2\theta - \sin^3\theta = 3\sin\theta\cos^2\theta - \sin^3\theta$; similarly, $\cos 3\theta = \cos^3\theta - 3\sin^2\theta\cos\theta$

49. $\dfrac{\cos\theta\tan\theta + \sin\theta}{\tan\theta} = \dfrac{\cos\theta(\sin\theta/\cos\theta) + \sin\theta}{\sin\theta/\cos\theta} = 2\cos\theta$

51. $\tan\theta + \cot\theta = \dfrac{\sin\theta}{\cos\theta} + \dfrac{\cos\theta}{\sin\theta} = \dfrac{\sin^2\theta + \cos^2\theta}{\sin\theta\cos\theta} = \dfrac{1}{\sin\theta\cos\theta} = \dfrac{2}{2\sin\theta\cos\theta} = \dfrac{2}{\sin 2\theta} = 2\csc 2\theta$

53. $\dfrac{\sin\theta + \cos 2\theta - 1}{\cos\theta - \sin 2\theta} = \dfrac{\sin\theta + (1 - 2\sin^2\theta) - 1}{\cos\theta - 2\sin\theta\cos\theta} = \dfrac{\sin\theta(1 - 2\sin\theta)}{\cos\theta(1 - 2\sin\theta)} = \tan\theta$

55. Using (47), $2\cos 2\theta \sin\theta = 2(1/2)[\sin(-\theta) + \sin 3\theta] = \sin 3\theta - \sin\theta$

57. $\tan(\theta/2) = \dfrac{\sin(\theta/2)}{\cos(\theta/2)} = \dfrac{2\sin(\theta/2)\cos(\theta/2)}{2\cos^2(\theta/2)} = \dfrac{\sin\theta}{1 + \cos\theta}$

59. From the figures, area $= \dfrac{1}{2}hc$ but $h = b\sin A$

so area $= \dfrac{1}{2}bc\sin A$. The formulas

area $= \dfrac{1}{2}ac\sin B$ and area $= \dfrac{1}{2}ab\sin C$

follow by drawing altitudes from vertices B and C, respectively.

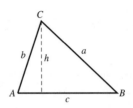

61. **(a)** $\sin(\pi/2 + \theta) = \sin(\pi/2)\cos\theta + \cos(\pi/2)\sin\theta = (1)\cos\theta + (0)\sin\theta = \cos\theta$
    (b) $\cos(\pi/2 + \theta) = \cos(\pi/2)\cos\theta - \sin(\pi/2)\sin\theta = (0)\cos\theta - (1)\sin\theta = -\sin\theta$
    (c) $\sin(3\pi/2 - \theta) = \sin(3\pi/2)\cos\theta - \cos(3\pi/2)\sin\theta = (-1)\cos\theta - (0)\sin\theta = -\cos\theta$
    (d) $\cos(3\pi/2 + \theta) = \cos(3\pi/2)\cos\theta - \sin(3\pi/2)\sin\theta = (0)\cos\theta - (-1)\sin\theta = \sin\theta$

63. **(a)** Add (34) and (36) to get $\sin(\alpha - \beta) + \sin(\alpha + \beta) = 2\sin\alpha\cos\beta$ so
    $\sin\alpha\cos\beta = (1/2)[\sin(\alpha - \beta) + \sin(\alpha + \beta)]$.

    (b) Subtract (35) from (37).   **(c)** Add (35) and (37).

65. $\sin\alpha + \sin(-\beta) = 2\sin\dfrac{\alpha - \beta}{2}\cos\dfrac{\alpha + \beta}{2}$, but $\sin(-\beta) = -\sin\beta$ so

$\sin\alpha - \sin\beta = 2\cos\dfrac{\alpha + \beta}{2}\sin\dfrac{\alpha - \beta}{2}$.

67. Consider the triangle having a, b, and d as sides. The angle formed by sides a and b is $\pi - \theta$ so
from the law of cosines, $d^2 = a^2 + b^2 - 2ab\cos(\pi - \theta) = a^2 + b^2 + 2ab\cos\theta$, $d = \sqrt{a^2 + b^2 + 2ab\cos\theta}$.

Solving Polynomial Equations

EXERCISE SET B

1. **(a)** $q(x) = x^2 + 4x + 2, r(x) = -11x + 6$
   (b) $q(x) = 2x^2 + 4, r(x) = 9$
   (c) $q(x) = x^3 - x^2 + 2x - 2, r(x) = 2x + 1$

3. **(a)** $q(x) = 3x^2 + 6x + 8, r(x) = 15$
   (b) $q(x) = x^3 - 5x^2 + 20x - 100, r(x) = 504$
   (c) $q(x) = x^4 + x^3 + x^2 + x + 1, r(x) = 0$

5.

x	0	1	-3	7
$p(x)$	-4	-3	101	5001

7. **(a)** $q(x) = x^2 + 6x + 13, r = 20$ **(b)** $q(x) = x^2 + 3x - 2, r = -4$

9. Assume $r = a/b$ a and b integers with $a > 0$:
   (a) b divides 1, $b = \pm 1$; a divides 24, $a = 1, 2, 3, 4, 6, 8, 12, 24$;
   the possible candidates are $\{\pm 1, \pm 2, \pm 3, \pm 4, \pm 6, \pm 8, \pm 12, \pm 24\}$
   (b) b divides 3 so $b = \pm 1, \pm 3$; a divides -10 so $a = 1, 2, 5, 10$;
   the possible candidates are $\{\pm 1, \pm 2, \pm 5, \pm 10, \pm 1/3, \pm 2/3, \pm 5/3, \pm 10/3\}$
   (c) b divides 1 so $b = \pm 1$; a divides 17 so $a = 1, 17$;
   the possible candidates are $\{\pm 1, \pm 17\}$

11. $(x + 1)(x - 1)(x - 2)$ 13. $(x + 3)^3(x + 1)$

15. $(x + 3)(x + 2)(x + 1)^2(x - 3)$ 17. -3 is the only real root.

19. $x = -2, -2/3, -1 \pm \sqrt{3}$ are the real roots.

21. $-2, 2, 3$ are the only real roots.

23. If $x - 1$ is a factor then $p(1) = 0$, so $k^2 - 7k + 10 = 0$, $k^2 - 7k + 10 = (k - 2)(k - 5)$, so $k = 2, 5$.

25. If the side of the cube is x then $x^2(x - 3) = 196$; the only real root of this equation is $x = 7$ cm.

27. Use the Factor Theorem with x as the variable and y as the constant c.
   (a) For any positive integer n the polynomial $x^n - y^n$ has $x = y$ as a root.
   (b) For any positive even integer n the polynomial $x^n - y^n$ has $x = -y$ as a root.
   (c) For any positive odd integer n the polynomial $x^n + y^n$ has $x = -y$ as a root.